RECEIVED

AUG - - 2019

NO LONGER PROPERTY OF
SEATTLE PUBLIC LIBRARY

RAINFOREST

D0311587

RECEIVED

AUG   2016

NO LONGER PROPERTY OF
SEATTLE PUBLIC LIBRARY

RAINFOREST

DISPATCHES FROM
EARTH'S MOST VITAL
FRONTLINES

TONY JUNIPER

COLOR IMAGES BY
THOMAS MARENT

ISLANDPRESS | Washington | Covelo | London

First published in Great Britain in 2018 by
Profile Books
3 Holford Yard, Bevin Way
London WC1X 9HD
www.profilebooks.com

Copyright © Tony Juniper 2019

Views expressed in this book are those of the author
alone and do not necessarily reflect the policies or perspectives
of any organizations that he is or has been affiliated with,
including the WWF.

All rights reserved under International and Pan-American
Copyright Conventions. No part of this book may be reproduced
in any form or by any means without permission in writing from
the publisher: Island Press, 2000 M Street, NW, Suite 650,
Washington, DC 20036

ISLAND PRESS is a trademark
of the Center for Resource Economics.

Library of Congress Control Number: 2019938252

All Island Press books are printed
on environmentally responsible materials.

Manufactured in the United States of America
10 9 8 7 6 5 4 3 2 1

Keywords: Amazon, Brazil, Congo Basin, Costa Rica, Friends
of the Earth, Greenpeace, indigenous communities, Rainforest
Action Network, World Bank, World Wildlife Fund, atmosphere,
biodiversity, biodiversity hotspots, climate change, cloud forests,
cocoa, deforestation, ecology, evolution, palm oil, pangolins,
poaching, rainforest, species diversity, temperate rainforests,
tigers, tropical rainforests, valuing nature

CONTENTS

INTRODUCTION **RAINFOREST MATTERS**   9

PART ONE: EARTH'S MOST VITAL SYSTEMS

1 RAINFOREST – A CLUE IN THE NAME   15
How tropical rainforests make clouds and recycle water,
sustaining farming far away from where they stand

2 LIVING ATMOSPHERE   37
How tropical rainforests lock up carbon and hold down global
temperature

3 THE ECOLOGICAL WEAVE   55
Tropical rainforests are complex ecological tapestries that work
through a multitude of intricate connections

4 EVOLUTIONARY TREASURES   76
The most diverse ecosystems on Earth, rainforests comprise
living assets of incalculable value

PART TWO: THE AMERICAS

5 THE 'NEW WORLD' RAINFOREST PEOPLES   106
European contact in the rainforests of the Americas decimated
complex societies and sustainable agriculture

6 FOREST CLEARANCE IN THE AMERICAS   125
The exploitation of 'virgin terrain' and the unfortunate mirage
of rainforest fertility

7 COUNTING THE COST: GREEN SHIELDS 138

The effects of deforestation range from droughts to increased
vulnerability to extreme storms – we are in an era of consequences

8 PROTEST AND SURVIVE 154

As awareness grew about rainforest destruction, campaigns
began to make some impact in protecting what was left

9 FOREST DIVIDENDS IN COSTA RICA 171

The destruction of the rainforests is not an inevitable result of
economic development – Costa Rica has shown otherwise

10 THE BEST FOREST CUSTODIANS 182

The most impressive forest outcomes can be gained through
the empowerment of indigenous peoples

11 TEMPERATE ZONE RAINFOREST 204

Rainforests also exist at cooler northern and southern latitudes
– they are distinct but connected to their tropical namesakes

PART THREE: AFRICA

12 LAST FRONTIERS: THE CONGO BASIN 220

The Congo Basin is the largest rainforest after the Amazon;
it is still largely intact but pressures are mounting

13 THE LOST FORESTS OF WEST AFRICA 234

The 'Upper Guinea' rainforests of West Africa are home to many
unique species, but their extent today is fragmentary

14 CLIMATE AND COCOA CHALLENGES 254

Conservation of the rainforests is far more likely to succeed
when smallholders are empowered to do sustainable farming

PART FOUR: ASIA AND THE PACIFIC

15 A SHORT TOUR OF THE EASTERN FORESTS 278

Explosive economic and population growth have taken a heavy
toll on Asian and Pacific rainforests

16 HOW TO DESTROY A RAINFOREST 297

Indonesia largely erased its rainforests over two decades, aided by the World Bank and the IMF – and multinationals

17 TIGERS, PANGOLINS AND HONEYCREEPERS 316

Hunting, forest fragmentation and the effects of introduced invasive species are all taking a toll on forest wildlife across Asia and the Pacific

18 POACHERS INTO GAMEKEEPERS? 331

As campaigners became effective at targeting companies causing deforestation, some began to change their practices

PART FIVE: WORTH MORE ALIVE THAN DEAD

19 GLOBALIZED DEFORESTATION . . . 345

Deforestation is driven by the global economy and policies to maximize growth through the export of natural resources

20 . . . AND GLOBAL SOLUTIONS 355

As the agenda started to change, informed by ecological science and economic reality, so the machinery of global government began to catch up

21 VALUING NATURE AND RAINFOREST 372

Companies and governments have increasingly come to accept the economic and practical value of intact natural systems

22 FUTURE FORESTS 380

New international agreements, extra money, company policies, public awareness and data-based tools have transformed forest prospects . . . but time is short

ENDNOTES 399

NOTE ON THE MAPS 425

PHOTO CREDITS 426

ACKNOWLEDGMENTS 427

INDEX 430

INTRODUCTION
RAINFOREST MATTERS

In the spring of 1990 Friends of the Earth placed an advertisement in *New Scientist* magazine inviting applications to lead the organization's tropical rainforests campaign. I'd done some campaign work with the International Council for Bird Preservation (now BirdLife International) trying to save the world's most threatened parrots from extinction, had a couple of science degrees and a passion since childhood for wildlife. It seemed like the job for me, so I applied. To my delight I was offered the role and six weeks later reported for duty.

Friends of the Earth operated out of a shabby office building in a then run-down part of Shoreditch, close to the City of London. It was a 1950s building, built on a V2 bombsite amid a row of Victorian warehouses. It struck me as a fitting location from which to wage the campaign for these wonderful ecosystems. First thing that Monday morning I met my new team to talk about strategy and where we needed to focus our efforts. There were boxes of papers everywhere, covering every conceivable subject linked with tropical forests. I was told how the archive was organized and how it was maintained. Then my new boss, Campaigns Director Andrew Lees, took me up onto the roof to tell me roughly what he had in mind.

Andrew was a charismatic and gritty figure, oozing fighting spirit. He smoked cigarettes and consumed a large mug of sweet black coffee while giving me a precis of the campaign to date, and where he thought it should head next. He wanted to raise political pressure for more action and that would come from a combination of research, media coverage, lobbying and public backing for our goals. These included ending unsustainable tropical timber imports to Europe, changing the policies of the World Bank (which had funded a lot of tropical forest destruction) and altering the practices of multinational companies, such as those in the oil and gas sector that were entering indigenous territories in the Amazon. There was also an ongoing campaign on Third World debt, which had been identified as one of the underlying economic drivers of the problem.

We'd need to raise the money to do all this, he explained, and build up the capacity of the Friends of the Earth International network, so that campaigners in the rainforest countries were better equipped for the epic struggle that lay ahead.

'Is that all?' I said. 'No,' replied Lees. 'But it's enough to be getting on with.' 'I'd better get started then,' I said, and so began my career as a campaigner for the tropical rainforests.

I've been at it ever since.

Back in 1990 awareness about the plight of the tropical rainforests was growing fast in relation to the threats to indigenous people and wildlife, especially across the Amazon. There, native societies were being driven into oblivion, their languages, culture and ways of life being lost forever, and so too were many species of animals and plants, including untold numbers still to be described, hemorrhaging into extinction right before our eyes.

There was far more limited awareness of the rainforest's hugely important role in combating climate change (a phrase then yet to reach general currency), through the absorption of carbon dioxide. A few months after I joined the team we issued a report on the scale of climate-changing emissions caused by deforestation. Having worked with leading climate change and deforestation experts, our estimate was that about a fifth of emissions were then being caused by tropical rainforest destruction—a figure confirmed later by better-resourced research agencies. The report got hardly a mention in the media.

And there was even less popular knowledge—or indeed much available scientific research—of the role of the forests in replenishing oxygen, or driving the circulation of fresh water, and how deforestation in South America might impact on the frequency and intensity of droughts in places as far afield as North America's Great Plains, Africa and Europe. In short, we didn't know the half of it. Rainforests were far more important than we had imagined.

What we did know, though, was that rainforests were already in crisis. The scale of forest loss in some regions was already extreme, in southeastern Brazil and West Africa, for example, and was indicative of what could soon lie ahead for other tropical rainforest regions. During the 1970s scientists had for the first time begun to compile comprehensive data sets on forest losses and in 1981 the UN published its findings in its first Tropical Forest Resources Assessment Project, detailing the huge scale of forest loss then taking place. In 1990 a more sophisticated analysis using satellite data estimated that the compound rate of tropical forest loss was running at nearly 1 percent per year.

That trend, if it remained static, would see the disappearance of all of the tropical forests within a century. There was a real sense of urgency, increasingly high stakes, and with that a huge responsibility to succeed. In the way of success, however, lay some very deep challenges. Not least of these were the rapidly growing populations of many rainforest countries and the related need, as many political and business leaders saw it, to expand economic growth so as to help ensure the expanding numbers of people didn't live in poverty.

Against this backdrop my Friends of the Earth predecessor Charles Secrett had in 1984 launched the world's first campaign to slow down and halt the destruction. In the pages that follow I tell the story of the battle for the rainforests, what I have seen at the frontlines, what has worked and how progress has sometimes been made. We start, though, with more on why these rainforest ecosystems are so important and extraordinary, and the crucial roles they play in the rise of truly twenty-first-century challenges—water security, climate change and conserving the Earth's staggering natural diversity.

Tony Juniper, Cambridge, 2018

PART ONE

EARTH'S MOST VITAL SYSTEMS

1 RAINFOREST–A CLUE IN THE NAME

How tropical rainforests make clouds and recycle water, sustaining farming far away from where they stand

If there is one factor that unites the diverse set of ecosystems we call tropical rainforests, then it is rain. This seems obvious. Rainforests take their name from the fact that a lot of rain falls upon them. But it is vital to know that they can also create it, and not only that, but help to move it far away from where the rainforests themselves are. The whole cycle is powered by photosynthesis, and how rainforests pump water into the air as part of the process of converting the sun's energy into chemical energy. In many rainforests it is possible to see them doing this on an almost daily basis.

I've observed this for myself a number of times, including on a trip to the Amazonian lowlands of Peru, sitting on a ridge looking toward the green expanse of a rainforest canopy close to the steep slope of the Andes. There I watched for a couple of hours the effects of the sun beating down on the sea of trees that stretched far into the distance. Birds flew back and forth, monkeys called and insects buzzed, and behind all of it a fundamental process took place. As solar radiation powered photosynthesis in the trillions of leaves held aloft on that canopy of trees, so they released more and more moisture until the sky above gradually turned from clear blue to hazy blue, and then grey. The moist air rose and condensed and the outline of trees on the distant higher slopes blurred, as cloud and forest seemed to become one. And this fusion between trees and atmospheric moisture was not just an impression. For when the clouds that evening discharged their loads of rain from beneath a dark crashing thunderstorm, the water would not only run down rivers in the rainforest, but also back up the stems of the trees, to emerge once more through those solar-powered leaves, and again flow into the atmosphere.

The first writer to suggest that there were connections between forests, the humidity of the atmosphere and climate was the German naturalist Alexander von Humboldt, who explored the Americas during the early 1800s. He saw rainforests while traveling along the Orinoco and the Casiquiare, the latter linking with the Rio Negro in the catchment of the Amazon, thereby connecting these two vast river systems.

At a time when knowledge was increasingly sought through more and more specific scientific disciplines, von Humboldt was a rare thinker who took the wider view, looking at the whole system as well as the individual species or components, such as trees or animals. He believed that the Earth was one great organism within which everything was connected and that the forest was a complex system, based not only on interactions between the many hundreds of thousands of distinct lifeforms that comprised it, but also with the atmosphere and water. He saw how forests released humidity into the air to form clouds, and then rain.

Despite von Humboldt's equatorial observations leading him to see the fundamental connections that exist between forests, clouds and rain, the phrase 'tropical rainforest' was first used only about a century later, in 1898, by a German botanist called

Humid air rises over a tropical rainforest, Tambopata-Candamo reserve, Amazon, Peru (TM)

Andreas Schimper. It took half a century more before Schimper's phrase (*tropischer Regenwald* in his native German) entered mainstream use, triggered by the publication of Paul Richards's 1952 book *The Tropical Rain Forest*. In the wake of Richards's work came a long-running debate among botanists, geographers and ecologists as to the definition of a rainforest. Generally speaking it was taken to be forest receiving more than 200 cm (80 inches) of rain or condensed mist per year, with that more or less spread throughout the year. Many humid tropical forests were included in the category, even though for some part of the year they had less rain or mist.

As time has gone by we've had more and more opportunities to gain a practical appreciation of how tight the connections between forests and rainfall often are. Take, for example, the 3,000-kilometer-long rabbit-proof fence built across the interior of Australia in 1907 to control rodents that had by then taken on plague proportions. On the side of the fence where the native vegetation remained, abundant cloud formed, whereas on the side with millions of rabbits, grazing animals and crops, the skies were clearer and there was less rain.

The recycling of water back to the air via vegetation is the result of a process called transpiration. On the underside of leaves are tiny pores that open to enable carbon dioxide to be extracted from the air. When plants do this, then through those same pores they lose water. As water is lost from leaves, so more is drawn up from the forest floor, through bundles of tiny tubes called xylem. These connect leaves all the way down to the microscopic hairs at the end of the roots.

There is no heart or other pump to push the water up and it moves from the soil and up to the leaves, including those on the ends of twigs, 50 meters up, via suction. As water is pulled up to the top of the tree via what are the equivalent of very long narrow straws, more is brought up from below, carrying the dissolved nutrients that are needed in the solar-powered manufacture of the molecules necessary for growth and reproduction. Water pressure inside the bundles of tubes is up to about fifteen times that of the atmosphere. When water reaches the tiny pores on the underside of leaves, the sudden reduction in pressure means the water can become a gas—water vapor.

Alongside water vapor the trees discard another gas—oxygen—a by-product of plants using sunshine to split water so as to get hold of the hydrogen they need to make sugars. Of all the oxygen released on Earth through photosynthesis, the tropical rainforests contribute about 20 to 30 percent of the total; most of the rest comes from photosynthetic plankton in the oceans.

Photosynthesis goes faster in brighter light, causing leaves to open their pores wider to permit the entry of more carbon dioxide while at the same time letting more water vapor out. Water also flows out more quickly when it's warmer, and when it's windier. So long as there is a lot of water, as there generally is in a tropical rainforest, the plants can avoid the wilting effects that accompany water scarcity.

Those thin green leaves to which most of us pay hardly a second thought are thus not only manufacturing complex organic molecules using sunshine, water and carbon dioxide caught from the atmosphere, but also releasing life-giving oxygen and water that makes new rain. When it comes to that latter function, the most extreme example of the synergy between vegetation and moisture is seen across the vast block of tropical rainforests that lie in and around the Amazon basin.

Amazon: Earth's largest freshwater system

Stretching for thousands of kilometers from the Atlantic Ocean on the eastern side of South America to the slopes of the Andes in the west, the Amazon basin is the world's largest freshwater system. Rain falling on the Andean mountains on the western side of South America is recycled several times after it first fell on the far side of the continent, where clouds roll in with moisture evaporated from the distant Atlantic. Even though it is in the tropics, some of the water falling on the high Andean slopes does not do so as rain, but as snow.

From December to May there is an annual melt that usually coincides with torrential rains over the forests. When this happens the depth of the Amazon's main river channel increases by between about 9 and 14 meters. Huge tributaries back up and vast areas are inundated, creating a particular kind of flooded rainforest called *várzea*. During February and March, when the water is at its deepest, it is possible to glide in a boat through the forest canopy. The huge slow-flowing swamp that is the result is the habitat of a range of specialist animals, including huge fish. During this wetter season, an area of forests bigger than New Mexico (around 350,000 square kilometers) is under deep water, with the main channel swelling to some 40 kilometers across. Even during the dry season the main river in its middle reaches is still about 11 kilometers wide.

The *várzea* is but one component in a complex system, whereby the trees and water work in concert. At over 6,000 kilometers from source to sea, the Amazon basin river system drains nearly half of South America, a catchment of more than 7 million square kilometers. From streams arising on the slopes of the Andes to the sparkling rivulets flowing beneath trees in the forests, the capillaries drain into rivers that eventually aggregate into massive veins such as the Rios Negro, Madeira, Purus, Tapajos, Xingu, Caquetá and Putumayo, which in turn fill the great vena cava of the Amazon itself, that each day discharges some 15 billion tonnes of freshwater—all of which first fell as rain—to the Atlantic Ocean.

During the course of a year the Amazon system moves about one fifth of all the freshwater traveling in all of the Earth's rivers. Where this vast continental river reaches the sea, its mouth is over 325 kilometers wide—greater than the distance between London and Paris—projecting a vast plume of freshwater hundreds of kilometers out into the Atlantic. So huge is the river that ocean-going ships can navigate inland to Peru—around two thirds of the way up toward its source. The Amazon system is connected from end to end not only by water and shipping, but also wildlife, including a species of catfish that makes an annual migration in the headwaters on Andean slopes all the way to its mouth and then back again.

The rivers will of course only maintain flow so long as the rain continues to fall from clouds. That in turn depends on a supply of new water getting into the air and to the slopes of the Andean mountains. As might be expected, the proportion of water entering the air from leaves, rather than that evaporated from the surface of the ocean, increases with distance from the Atlantic, and with increasing distance from the sea, so the importance of the forest in maintaining rainfall goes up.

Why the rainforest is like a green ocean

In the tropical rainforests, where the organic green engine of photosynthesis hums at high speed, a single large tree can pump up over a thousand liters of water per day from the soil to the air, and in so doing dramatically changes the humidity above the forest where it grows. The vapor is lighter than the rest of the air and this triggers convection currents that take it skyward. At higher altitudes the air is cooler and this causes the vapor to turn back to tiny liquid droplets via condensation. The return to the liquid phase requires not only lower temperature, however, but also a non-gas surface upon which the return to liquid can be precipitated. You can see this taking place on a cool window in a steamy kitchen, or on grass at the dawn of a new day that follows a warm but cloudless summer evening. There are no windows or grass stems high above the rainforests, though, so other surfaces facilitate the return to liquid.

Whatever gets up to the cooler altitudes, where condensation can most readily take place, will of course be very small, and also very abundant. The materials that do this take the form of tiny floating non-gas particles called aerosols. These are the microscopic nuclei upon which water vapor condenses to create the tiny water droplets that come together in clouds. There are many different sources of these cloud-seeding materials, including dust kicked up from storms over deserts, salt grains from sea spray and various organic materials. Among the latter are compounds released through those pores on the leaves themselves, including terpene and isoprene.

At millionths or even billionths of a meter across, numberless trillions of these minuscule particles drift up from the forest in the humid air and in the presence of sunshine unite with oxygen to form very fine dust particles that have an affinity for water. These are among the tiny surfaces upon which condensation occurs. This tree-generated cloud fuel is augmented by billions of trillions of pollen grains rising into the air every day from the vast number of flowers, as well as fungal and bacterial spores. The result is a vast bank of cloud-seeding material being released by the rainforest itself, magic dust feeding into rising moist air to generate cloud, and then rain—lots of it.

So it is that the Sun's energy is harnessed by plants to lift water into the atmosphere, uniting with microscopic particles to create the rain that replenishes the headwaters of rivers, refills the aquifers that give rise to springs and moisten the soils that enable plant growth, and on a colossal scale. Each day some 20 billion tonnes of water vapor is emitted from the Amazon basin rainforests—more even than the huge quantity of liquid water that flows each day along the Amazon river itself. That vapor is a globally significant source of freshwater and a major proportion of the 90 percent of all the water that reaches the atmosphere via plant transpiration on land each day. The fact that only about 10 percent of the water arriving in the air from land is via simple evaporation (in other words isn't mediated by plants, and instead is coming directly from soils or concrete) underlines the vital roles played by vegetation in recycling freshwater.

Having said this, it is important to remember that most of the water vapor that goes into the air is from the surface of lakes and oceans (which are of course pure water). The oceans are more extensive than the land, but surfaces covered with forests can evaporate at least as much water as an equivalent area of lake or ocean. Dense stands of trees with understories of shrubs and covered in epiphytes present multiple layers of vapor, emitting surfaces, whereas the sea or a lake has just one.

Recent research suggests that strong parallels can be drawn between how rainforests and oceans work. One study using images generated by cameras borne on satellites and aircraft found that Amazonian clouds have a striking resemblance to maritime clouds, leading some researchers to describe the rainforest as a 'green ocean.' The parallel is seen in the expansive forest surface covering millions of square kilometers, stretched out beneath the atmosphere, vast, wet and exposed to the winds that move moisture, very much like an actual ocean.

How the rainforests pump water around the planet

Significant areas of the green oceans of rainforests that once covered so much of the humid tropics have, however, been drained, and (as we'll see later) with increasingly serious implications for humans. This is because the rainforests not only replenish cloud and thus freshwater, but, it seems, also help drive large-scale air movements, effectively pumping moisture-laden air inland. This is the result of the condensation of the vapor released by the forest, which causes dense clouds to form above the sea of trees. The condensation leads to a drop in air pressure, causing warm moist air to be sucked in from over the ocean. This brings with it the airborne water that creates the rains that moisten the forests in the first place—and which is then recycled by the forest into new clouds.

Clouds condense over the forests of the Peruvian Amazon (TJ)

One person who has spent a lot of time looking at how this might work is a British scientist called Peter Bunyard. I've known Peter for years and bumped into him in 2011 at a conference on the conservation of tropical forests in Colombia. He has spent much of his career researching and writing about the Amazon rainforests and he excitedly told me about new experiments he was planning that should help explain what may be a very powerful force created by water and trees working in synergy. At the time Bunyard was researching at Sergio Arboleda University in Bogotá and had become absorbed with understanding how forests were apparently driving atmospheric flow through a previously unexplained mechanism.

The inspiration for Bunyard's experiments came from two Russian mathematicians from the Institute of Nuclear Physics in St. Petersburg: Anastassia Makarieva and Viktor Gorshkov. These researchers had put forward an idea called the 'biotic pump theory.' This explained how the physics of condensation would lead to surface airflows because of a force that enabled living systems to literally pump air. Without it, the humid ocean-derived airflow would dwindle away, leading to less and less rain. Through this theory it was possible to see how the endpoint of widespread deforestation could be desertification.

Bunyard explained to me how the mechanism by which this is believed to work hinges on how, when water goes from vapor to droplets via condensation, latent heat is released. The release of that heat during condensation results in upward airflow, giving impetus to the motion of the column of air above the rainforest—more than would be the case than by simple convection alone (the effect of warm air rising from the heating land). The biotic pump theory says that on top of the release of latent heat (and convection) an additional effect is created when vapor turns back to water droplets. This is because, during condensation, when water goes from a gas (vapor) to liquid (droplets), it occupies a smaller volume, causing the air to collapse inward, thereby creating more momentum behind the rising air. This idea only works, however, if the air drawn in by condensation is coming from *beneath*, because that is where the air pressure would be higher. It is on this basic point that some scientists disagree, saying that condensation draws in air from *all* directions and, if that is the case, then the biotic pump theory doesn't hold, as it wouldn't be producing as much upward motion.

Instead of simply going back and forth with an exchange of theories, Bunyard decided to conduct experiments to measure what happens when condensation occurs, and whether it could indeed actually drive air flow. At his home in Cornwall, fittingly a converted watermill, he spent years refining a method to try and measure the biotic pump effect. A series of air-filled tunnels fitted with various pipes, refrigeration elements, air pressure sensors, an ultrasonic anemometer, thermocouple, relative humidity hygrometer, heating mat and 'rain' collector were combined to create the ability to measure how condensation might impact on airflow. Bunyard's work did indeed reveal a relationship and showed how, as water goes from gas to liquid, a powerful directional force can be created.

'It appears my experiments are the only ones that have ever been done on this process,' he told me. 'Nobody has ever done them before and I say the biotic pump theory is correct; it helps confer the upward movement which comes from this rarefaction of the air brought about by condensation.' The implications of this are truly profound, for it might confirm how the tropical rainforests are having an impact on atmospheric circulation at a far larger scale than previously imagined.

As these atmospheric forces over the forest push the air to higher altitude air pressure is reduced at lower levels, as the vacuum effect of the biotic pump pulls air skyward, and so more air is drawn in from the side to maintain the circulation, which in the eastern Amazon is moist air flowing in from over the Atlantic Ocean. The Russian researchers calculated the strength of the biotic pump and found it to be vast, powerful enough to be driving not only regional winds from the ocean to forest but also the Hadley Cells that are at the heart of the Earth's atmospheric circulation.

The Hadley cells are comprised of air rising near to the equator that when it reaches high altitude flows north and south, falling again in the subtropical latitudes. When the air is driven up over the forests it forms clouds and drops the moisture it carried as rain, so that the air reaching the surface again, north and south, is very dry. This is why the equatorial rainforest belt is mostly bordered by deserts—the Atacama and Sonoran deserts in the Americas, the Sahara, Namib and Kalahari deserts in Africa, and the Gobi and Australian deserts in the Eastern Hemisphere. Having descended over the dry subtropics, the air then returns toward the equator at the surface, where the movement is manifest in the trade winds that run over the ocean, again picking up masses of moisture.

Bunyard's work on the ideas put forward by the Russian scientists not only concerned what was going on above the forest, but also inside it. 'The forest canopy maintains a high relative humidity beneath it during the course of the day, which helps soils retain their moisture rather than being dried out from being exposed to sunlight, and that's very important to appreciate,' he explained. 'A full closed canopy rainforest, with its multiplicity of different trees and vegetation and the different layers, provides a constant source of water to be transpired. In other words, if you put a plantation instead of the rainforest it will not serve the same role; it won't prevent soils drying out, or work in the same way.'

So, while it was previously believed that it was warm temperatures that drove rising air that sucked in the moist winds from the oceans, evidence increasingly suggests that evaporation and condensation are also powerful factors in their own right. Thus a wet forested region that can evaporate much more water per square kilometer than an adjacent ocean—and which via condensation creates a lot of cloud—draws moisture-laden air toward it. The condensation is also in large part facilitated by the forest, through that flow of cloud-seeding dust. Thus when forests are removed, there is less evaporation over the land compared with the sea, and the wind direction can be reversed, going from the land to sea. In the long term that will result in a drier forest, grassland or even desert.

Bunyard was keen to emphasize to me the importance of the biotic pump idea in understanding this: 'Climatologists have so far excluded the biotic pump from their climate models, sticking to the notion that temperature differences are the driving force of the Hadley circulation. According to the current models deforestation will result in a 12 to 15 percent reduction in rainfall over the western Amazon, whereas, on theoretical grounds, at least, the inclusion of the biotic pump will result in a 99 percent reduction in rainfall in the western Amazon as a consequence of deforestation. That's why I was determined to investigate the actual physics of condensation.'

The drying effects of deforestation are already apparent. Whereas the Amazon used to be wet year-round, albeit less wet during some months, there is now a regular drier season and the length of the wetter period is diminishing. One study found that since 2000 some 69 percent of the Amazon rainforest received less rain than before. Deforestation and degradation of the rainforests is undoubtedly a factor. By 2013 the total degraded area of Amazon forest may have hit more than 1.25 million square kilometers. When added to the area completely cleared, to make way for pastures and soya fields, then the total may be over 2 million square kilometers. The Brazilian Amazon rainforests account for nearly half of this figure.

Less condensation occurring over the forests because of fewer trees not only means that there is less water to circulate within the Amazon basin, but also less that can be exported outside of it, including to Brazil's southeast, a vital agricultural region that requires a lot of water to keep it going.

Sky rivers and the long-distance movement of water

Looking across the world at the distribution of the subtropical deserts you might expect there to be one at the latitude of São

Paulo in Brazil, covering what is instead the vast agricultural area of Central South America centered on the Rio de la Plata basin. This productive region, located at about the same latitude as the Kalahari and Atacama deserts, is bounded by the Atlantic Ocean to the east and, crucially, by the Andes to the west. In the Hadley circulation it might be expected to be an arid zone, but it is relatively moist. That's because it is being fed with aerial flows of water vapor supplied by the Amazon rainforests via the biotic pump effect, generating not only the vertical winds that spread north and south but also lower-level airflows. These conveyors take moisture to destinations far from where it was first evaporated from leafy rainforest canopies.

Rain falling in Central South America is in large part derived from the Amazon forests to the north, with water vapor being transported at low level and guided by the 6-kilometer-high barrier of the Andean mountains. Such long distance aerial water flows have been dubbed 'sky rivers' and are the reason it's possible to have such productive farming in the center of South America, south to Paraguay and northern Argentina. On the far side of the Andes and on the other side of the Atlantic Ocean, there are deserts.

The Amazon rainforest might thus be seen as a beating heart in the freshwater system, pumping air as well as water, driving arterial vapor flows with venous rivers eventually taking water back to the oceans from whence it originated. Over the tropical Atlantic the powerful sunshine drives more evaporation, which the heart then pumps back over the forest, where cloud is seeded by a column of organic dust created by the trees themselves. It is a beautiful and remarkable system, working simultaneously at microscopic and planetary levels and running on intricacies and interdependencies so finely poised as to defy belief.

As we've begun to understand more about how the forest–rain relationships work, so we've also started to better appreciate the implications of deforestation for rainfall. Deborah Lawrence and Karen Vandecar are two American scientists who've spent a lot of time looking at this, including at planetary scale. Using a wide range of modeling exercises they've investigated the extent to which rainforests are not only sustaining rainfall close to where they stand, but also very far away.

I spoke to Lawrence and asked about their research. 'I looked at the results of many different modeling exercises in which tropical forests were eliminated to see how the climate system would respond. In study after study you could see that, although the deforestation was happening in one tropical zone or another, or across the entire tropics, there were changes in rainfall that were very remote—as far north as the American Pacific northwest, the UK, China, parts of Asia; they impacted laterally, too, with Amazon clearances having effects in Africa.'

Models can't predict precisely what will happen in any particular case, but they can tell us something about the risks that come with the choices societies make, including replacing tropical rainforests with crops, pastures, cities and mines. Lawrence told me that, using her models, 'I started understanding that tropical forests have a huge role not only in regulating local climate but also global climatic conditions. It's really down to how they change the movement of energy by evaporating water from the surface of the leaves. That changes everything.'

I asked if the research she'd been involved with had identified regions at particular risk. 'It looks like the Midwest is quite at risk, there's no doubt about that. This is a big breadbasket and I'm looking at all my triangles marking the places where there's a negative impact and they just make a corridor north from Texas.' That corridor is the Great Plains, the vast undulating landscape

Rainforest rivers are sustained by trees. San Fernando Waterfall, Costa Rica (TM)

that runs through the center of the USA and into Canada, and where a great deal of the world's food is grown.

Lawrence and Vandecar have presented evidence to show that the impact on rainfall from deforestation is not simply a theoretical future risk, but already happening. For example, in Thailand the beginning of the dry season is drier than usual, probably because of deforestation. In the Amazon, too, they linked the smoking gun of deforestation with progressively drier conditions. One important question they've investigated is at broadly what point water transfer systems might collapse. As is the case with other complex multi-factor modeling, it is difficult to be precise, but their analysis suggests that in the Amazon, and maybe the Congo Basin too, the tipping point beyond which the system changes from one state to another may be between 30 to 50 percent deforestation.

Progressive drying caused by damage to the forests won't only affect regional and long-distance water transfer, but also the viability of the rainforest that remains, as, for example, fire risk goes up with drier conditions. In other words, beyond a certain point, areas of rainforest might enter a spiral of drying that eventually leads to its transformation into savanna, grassland or even desert, in turn altering rainfall patterns and impacting on farming, thousands of kilometers away. As might be expected, given their findings, Lawrence and Vandecar suggest that it would be wise to retain large areas of tropical rainforest and to avoid large-scale deforestation in any single location.

While their research was focused mainly on large blocks of forest that release a lot of water vapor, there are other forests that more commonly occur in narrow strips along the sides of mountains. These also play important roles for water security, but more through how vegetation harvests moisture from the air rather than their role in evaporating it.

Filling rivers with mist: how cloud forests work

Mist mingling with leaves, stems, fronds and mosses can make for very wet forests, taking moisture from clouds even during the dry season, enabling year-long river flow even in the absence of actual rain. From Costa Rica to Vietnam and from Colombia to Indonesia, cloud forests, with their cool upland conditions, support stands of trees festooned with a myriad of other plants, including mosses, lichens, ferns, bromeliads and orchids. There are even the seedlings of other trees growing on the branches and trunks of bigger trees. These thick cloaks of vegetation, and the huge combined surface area of leaves, stems and fronds, is the mechanism whereby the condensation of clouds causes water to be extracted upon slopes where each day dense pervasive mist forms.

The Los Angeles Cloud Forest in Costa Rica lies at about 1,125 meters above sea level and is fed by a near constant flow of cloud that builds up when the maritime air coming from the warm Caribbean is forced to higher altitudes, leading to the generation of mist. The moisture loaded into the warm maritime air is augmented by evaporation from the lowland rainforests it passes over before the saturated flow envelops hillsides to create some of the most amazing forests on Earth.

In this subtropical zone, as in all dense rainforest, the name of the game for plants is to get to the Sun and every strategy is employed to do that. On a trip there in 2016 I walked beneath strangler figs that smothered host trees upon which they were first seeded, enveloping the trunk and branches beneath its ever-expanding form. The fig, in its turn, was festooned with mosses, ferns and bromeliads. In this cloud forest alone there were some 200 species of bromeliads, perched aloft on tree branches (or fence posts around nearby farms), each one creating an ecosystem of its own, with the pools of water contained in their

This tree in the Costa Rican cloud forest was making its own stream from mist (TJ)

whorl-like structures vital for the breeding cycle of many species of frogs and insects. Some bromeliads can hold up to 4 liters of water. Monkeys, toucans and snakes drink from them.

Great drips of water fell from the canopy, creating loud wet thwack sounds as they hit the big floppy leaves of the plants in the ground layer. On the forest floor, where the endless drips create saturated conditions, grew a multitude of palms and ferns.

Some of the ferns were as big as trees, including monkey-tailed tree ferns that reached 20 meters in height.

Underneath some of the bigger trees the fall of water was like rain. One gigantic tree generated its own stream, as moisture concentrated from mist ran down its branches and trunk to create a flow of water beneath the sprawling epiphyte-laden crown. Some of the epiphytes that hung from its high branches had been taken up there as seeds by birds and since they germinated had grown long trailing roots that dangled down to the ground some 20 meters beneath. These aerial roots were themselves covered with mosses that also ran with water.

It is in these wet upland forests that rivers are born and fed, and in many tropical countries such ecosystems are vital for water supply. For example, parts of the Western Ghats of India are clothed with cloud forests that contribute water to the dozens of rivers that originate in those mountains and which provide drinking water, irrigation and power from hydroelectric dams for approximately 245 million people. India's fast-growing population and economy are, of course, 100 percent reliant on freshwater, so the forests are a major economic resource. Whether it is extensive lowland forests exporting the moisture that generates rain, or cool mountain forests that are harvesting water from mist and cloud, the linkages between forests and water are fundamental.

Reflecting heat back into space: albedo

The set of connections between forest loss and water transfer is not the only aspect of concern in how the continuing destruction of the rainforests might affect outcomes for human societies and economies. This is because the world can also expect to be buffeted by the ill winds of volatility arising from rising temperatures.

Deborah Lawrence says that, unlike forests in the temperate and boreal zones, the tropical rainforests actually reflect heat back into space—a so-called albedo effect; removing the forests in these equatorial latitudes causes the land to absorb more heat, elevating air temperatures. This is in addition to the removal of the cooling effect of evaporation from plants, meaning that the clearance of tropical rainforests is almost always accompanied by temperature increase. This effectively makes rainfall scarcer, because higher temperatures cause what moisture *does* fall to evaporate more quickly, leaving less available for crops and everything else.

The bioclimatic effects of rainforest loss are thus several-fold, causing changes to local rainfall patterns, impact on long-distance water transfer, changing atmospheric circulation and causing local and regional temperature increase. Unfortunately, these are not the only climatic consequences of deforestation, for there is a further, hugely significant global linkage between forests and climate: the impact of the rainforests in absorbing and emitting carbon dioxide.

2 LIVING ATMOSPHERE

How tropical rainforests lock up carbon and hold down global temperature

When Friends of the Earth issued its first report on climate change and tropical rainforests in 1990 we highlighted the connection between deforestation and carbon dioxide emissions. Despite our best efforts, however, it took the best part of twenty years for rainforest carbon to get firmly on the agenda. Even now, its impact is not widely known, outside expert circles. Yet rainforests are among the most important of the Earth's carbon stores and their preservation and restoration is one of the least expensive actions that we can take to reduce carbon dioxide emissions, remove carbon from the atmosphere, and in the process help avoid the worst effects of climate change.

An extraordinary record of the fluctuating concentration of carbon dioxide in the Earth's atmosphere can be observed at the British Antarctic Survey (BAS) headquarters in Cambridge. Here, in rooms kept at a constant temperature of minus 20ºC, is where one of the most remarkable archives ever to have been compiled is maintained: cylinder-shaped lengths of ice drilled from east Antarctica—the thickest ice in existence today. They're translucent white, about 1 meter long, 15 centimeters in diameter, and record changes in the Earth's climate going back to times before modern humans evolved.

This Antarctic ice first fell as snow, and as successive layers built up on top of one another it became compacted and today forms ice up to 5 kilometers deep. The further the scientists have drilled down, the older the ice that they've brought up. At the bottom it's almost a million years old. Trapped within each sample are tiny bubbles of air, literally frozen in time from when the snow fell. BAS scientists have analyzed these samples of ancient atmosphere in order to help understand changes that took place over millennia.

I sometimes visit this ice archive as part of my work with the University of Cambridge Institute for Sustainability Leadership, working with leaders from different organizations to help them better understand the profound changes that human activities are causing to the Earth's atmosphere. As you hold the disc-shaped wafers of ice that have been frozen for thousands of years, your body warmth causes it to melt, and as it returns to liquid it's possible to hear a crackling sound as the gases that have been held tight for so long are reunited with the air.

We know from these ice cores and other evidence that our planet's atmosphere has been more or less able to trap heat depending on the concentration of greenhouse gases, particularly the quantity of carbon dioxide. By analyzing the air locked inside those ice samples, BAS scientists have been able to draw a graph that reveals the changing levels of carbon dioxide going back more than 800,000 years. Over that period carbon dioxide levels have fluctuated between about 180 and 280 parts per million, with the highs and lows matching the peaks and troughs of temperature changes that took place during the Pleistocene 'Ice Ages' and the warmer interglacial periods that separated them.

Toward the end of the time series, in the period that followed industrialization in the eighteenth century, there's a dramatic change: the concentration of carbon dioxide goes up, fast, and on an ever-steeper trajectory. Direct measurements of the modern atmosphere reveal how since the late 1950s carbon dioxide concentration has gone up very quickly, in 2013 passing, for the first time in at least 800,000 years (and very likely several million years), the landmark of 400 parts per million. That huge and continuing upswing—the principal reason for global warming—is being caused mainly by the combustion of fossil fuels, soil degradation and deforestation. The consequences are already manifest in the repeated smashing of temperature records, the melting of glaciers, ice caps and sea ice, rising sea levels, more extreme weather, including droughts, floods and heatwaves, more frequent severe storms and alterations to seasonal patterns. And, of course, the impacts are expected to become more serious as the years unfold. The question is whether we can do enough in time to avoid potentially disastrous outcomes.

What might be done hardly needs restating: a switch from fossil fuels to low carbon alternatives, more efficient energy use and the expansion of energy storage technologies are among the big shifts needed. All of that is beginning to occur, at least in some parts of the world. Less prominent, however, has been the question of how to stop and reverse deforestation, for the clearance of tropical rainforests is a cause of massive emissions of carbon dioxide.

The rainforest carbon store

In past centuries, deforestation happened mainly in the temperate regions. But today forest loss occurs mainly in the tropics, where rainforests take carbon dioxide from the atmosphere as part of their chemical feedstock to manufacture complex molecules used in making leaves, flowers, fruits, seeds, twigs, branches and trunks. This plant biomass is thus a store of carbon that would otherwise be in the atmosphere as carbon dioxide. As forests grow to maturity, so the store of carbon they hold increases, such as in the trunks of massive trees. In some forests the carbon

Land cleared for a forestry plantation reveals the vast body of peat that previously lay beneath the rainforest. Riau Province, Sumatra (TJ)

store is also significantly increased through the accumulation of unrotted plant remains.

Considering the very rapid breakdown and recycling of organic material in a tropical rainforest (more on which later), it is perhaps paradoxical that it is the *lack* of decomposition that has led to the formation beneath some forests of vast peat swamps. Peat forms beneath the tropical rainforests, as elsewhere, when the flow of dead organic material coming from the vegetation growing on top of it exceeds the rate at which it is being broken down. The slowdown in decomposition occurs under acidic waterlogged conditions in which microbiological and fungal activity is very low or absent. Such conditions have over millennia created huge peat swamp forests across expanses of the Southeast Asian islands of Sumatra and Borneo. Another massive body of peat, holding an estimated 30 billion tonnes of carbon, lies beneath the Congo Basin rainforests, and was discovered only recently. Our understanding of these systems is still very much work in progress.

The peat that is such an important feature of large landscapes across Borneo and Sumatra is made from thousands of years of tropical photosynthesis. In some places it is more than 20 meters thick and from these peaty landscapes the rivers flow dark brown, or even black, as organic compounds leach from the unrotted plants and into the waterways. It is estimated that about 800 tonnes of carbon dioxide is released from the clearance of one hectare (that's 100 meters by 100 meters) of previously undisturbed forest. However, the destruction of tropical rainforests growing on peat causes much bigger releases of greenhouse gas pollution than those on mineral soils, which are comprised mainly of weathered rocks.

Overall, it is reckoned that about 230 billion tonnes of carbon is tied up in tropical forests—equivalent to about a third of all the carbon stored in all the economically recoverable oil, gas and coal reserves. Around 80 billion tonnes is located in just two countries: Indonesia and Brazil. Those two nations are also the highest emitters of carbon dioxide from deforestation and in 2011 were sixth and seventh in the world ranking of carbon polluters, ahead of major industrialized countries such as Japan and Canada.

Since 1850 at least a quarter of carbon dioxide pollution deforestation and today it is estimated that between 14 and 21 percent of annual emissions arise from forest damage, the vast majority in the tropics. The wide estimate range reflects the difficulties that accompany the measurement of carbon entering the atmosphere because of land use changes, especially from disturbed and damaged soils and forest degradation. While it is relatively straightforward to know how much coal has been fed into a power station, and thus the carbon dioxide being released in flue gases when it is burned, it is far more difficult to know precisely how much is being released from, for example, soils beneath a forest that has been damaged by logging.

Better data gathered from satellites and aircraft will in future give a more accurate picture but we can be confident nonetheless that the rising carbon dioxide concentration in the atmosphere is in large part down to how we are changing forests. Most models suggest that this contribution is larger than all global transport emissions put together (including airplanes). And the effects of this carbon pollution are additional to those from disturbances to the forest bioclimate described in the last chapter, which has such a huge bearing on local temperature and the water cycle.

Two degrees: a global warming danger threshold

The more carbon dioxide that is released into the atmosphere from the destruction and damage to forests, then the more difficult it will become to hold the overall average global temperature increase to within the globally agreed limit of two degrees centigrade, compared with the pre-industrial average. How best to do this is a vitally urgent question, because the loss of the rainforests presents a double whammy for the climate, for rainforest ecosystems are not only storing carbon but also removing carbon dioxide from the atmosphere, endlessly topping up those stores, on a massive scale. Even mature old forests that seem like they have reached a state of equilibrium absorb huge quantities of carbon dioxide from the air, as vast canopy trees grow thicker trunks and as deeper layers of peat accumulate in the soils. When the forests are set ablaze or cleared to make way for plantations, we are thus not only releasing carbon stores but destroying what are also among the world's most effective carbon-catching systems.

The carbon situation is made all the more challenging because in recent decades the tropical rainforests have been sucking up carbon dioxide ever more rapidly, as plants have responded to increased concentrations of the gas by growing faster. One study reported that global concentrations of carbon dioxide would be 85 parts per million higher today were it not for enhanced vegetation growth. Overall, it is estimated that intact and recovering tropical forests are removing between 1.2 and 1.8 billion tonnes of carbon each year, which is about 10–15 percent of what we need to achieve to meet the two-degree target. If deforestation and forest disturbance were stopped, including from logging natural forests, and emissions from that source closed off, and if large-scale forest restoration were undertaken, then the tropical forests could account for a third of what we need to do to meet that two-degree target. That is a huge opportunity, if only we can find the means to grasp it. We'll come back to how we might do that later on but in the meantime it is vital to signal a sense, not only of opportunity, but also urgency.

Feedback: how forests could create more warming

The urgency arises not only from the narrow and closing window for action in limiting emissions to avoid a two-degree temperature increase, but to manage the growing risk that climate change impacts will cause the rainforests to dry out, turning them into savanna woodlands, grasslands or even deserts. The progressive drying of some forest areas is already

A relatively small proportion of big trees hold a high proportion of rainforest carbon. Western Ghats, India (TJ)

apparent, caused not only by damage to the biotic pump effect that moves water vapor from over the oceans and toward the forests, but also because rising carbon dioxide concentrations are changing climatic patterns.

Take, for example, the effect of rising sea surface temperatures that accompany the El Niño warming that periodically takes place in the tropical Pacific Ocean. Drought conditions in Borneo and Sumatra are associated with strong El Niño conditions and during such an episode in 2015–16 huge rainforest fires broke out, causing massive carbon dioxide emissions, as tracts of peat swamp forest went up in smoke. The extent to which particular events like this have been caused by climate-changing carbon pollution is a matter of ongoing debate, although recent modeling presents evidence for a likely future doubling in the occurrences of severe El Niño conditions due to warming caused by human impacts on the climate and ocean systems. This would result in major consequences for forest ecosystems, as recent drought and fire events in Indonesia have so vividly revealed.

The Amazon might also be affected by warmer sea surface temperatures in the tropical Atlantic, as a result of a shift in the intertropical convergence zone reducing the quantity of water vapor flowing over the forests from the sea. A severe Amazonian drought in 2005 and another in 2010 were both associated with high sea surface temperatures in the Atlantic.

Climate change models can't predict the future, nothing can, but as is the case with the modeling of water flows they can tell us something about risk, including so-called feedbacks. Take for example, what happened during the Amazon drought of 2010. Under normal circumstances the rainforests absorb carbon as photosynthesis extracts carbon dioxide from the atmosphere to make new plant material. But as a result of exceptionally hot and dry conditions the plant mass was reduced and is estimated to have led to the *release* of about 2 billion tonnes of carbon. This is roughly equivalent to China's combined emissions from fossil fuels and cement production for that year. And it was caused by a heatwave affecting one area of tropical rainforest.

There is evidence that severe heat is increasing in the Amazon and that climate change will make this worse as the world warms up. Extreme heat comes on top of the effects of the deforestation that is leading to reduced rainfall and the two things together could cause the forest to shrink and shed carbon into the atmosphere, adding to the effects of that released from fossil fuels and other sources. This is called a positive feedback, whereby more warming leads to more emissions that lead to more warming. At its worst, it could turn into a cycle that is beyond anything we humans might do to stop it. This is one reason why the two-degree limit is considered so important, for if we go beyond that then it is feared that our ability to avert the potentially far more serious climate changes that could follow would be severely compromised.

Remarkably, however, some rainforests seem to have the means to respond to stress arising from hot and dry conditions, at least up to a point. Under drought conditions it might be expected that photosynthesis would slow down, so that the rate of evaporation is reduced, enabling trees to conserve water. What surprised scientists looking at forests suffering from drought was how the biotic pump effect that has such profound influence in regional climate and water transfer appears to have the means not only to withstand but even to reverse droughts that periodically affect tropical rainforests.

Evidence from the Amazon suggests that in response to dry conditions the rainforests release more water vapor. Even when rainfall is low, they pull up water from underground and produce more condensation above the forest, thus powering the biotic pump that drives the conveyor of moist air from the sea. In other words, it seems that some forests have an inbuilt mechanism to draw in moist air when they're affected by drought.

Degradation and fragmentation

This amazing ability of rainforests to respond to drought, and their ability to hold and catch carbon, is of course dependent on how much forest cover remains—and its condition. Logging operations can severely degrade rainforest, as the large trees generally targeted by timber traders play such a major role in carbon storage, accounting for between 25 and 40 percent of above-ground plant material, despite constituting only about 1–4 percent of trees above 10 centimeters in diameter. Should those trees be converted into structural timber, window frames or good-quality furniture, then some of the carbon they contained will continue to be locked up, but most often logging is not a sustainable activity that permits renewed tree growth but the first step in total forest clearance. Even when forest cover remains following forestry operations, degradation of forests by logging and other causes of disturbance, the carbon density in the forest can be drastically reduced. Indeed, one 2017 study estimated disturbance and degradation to be the source of about *two thirds* of tropical forest carbon emissions, considerably larger than outright deforestation.

Most of us are familiar with images of forests being logged, cleared or burnt, but less so with the fact that they are also being split into a multitude of patches, often quite small, and often degraded. Fragmentation of forests by roads, farms, fire and logging is a further factor increasing emissions, as the creation of more forest edges reduces the quantity of carbon trapped.

A 2017 estimate, based on satellite images, suggests that across the tropical forests as a whole what were once large contiguous blocks of natural vegetation are now fragmented into about 50 million separate pieces. The more the forests are fragmented, the longer the length of forest edge that exists, to the point now where there are some 50 million kilometers of it. So-called 'edge effects' include increased wind and desiccation causing higher tree mortality, thus increasing carbon emissions for long after a forest fragment was created. Edge effects can change the forest microclimate up to 300 meters into the forest, meaning that in small forest fragments 600 meters across the effects of fragmentation can penetrate across all that is left.

Intact and healthy forests not only keep climate-changing emissions out of our planet's atmosphere but also help protect life and property against the impacts of the more extreme conditions that are coming in the wake of climate change. Right across the tropical rainforest countries deforestation has exposed people to increased risks from the kinds of extreme weather that accompanies the climate changes caused in part by deforestation. It's another double whammy, where the loss of forests leads to more extreme conditions while at the same time rendering people more vulnerable to them. From the Philippines to El Salvador and from Sierra Leone to Venezuela, recent decades have seen extreme storms cause massive damage to life and property, with, in many cases, the effects made worse through the loss of tree cover.

Forests and healthy soils beneath them hold water and thus moderate the flow of rain into rivers and thus can smooth flood peaks, helping to reduce the severity of extreme events. With trees removed, heavy rain flows from the land more quickly, taking soils and sometimes bringing catastrophic consequences.

Anastassia Makarieva and Victor Gorshkov, the scientists who first proposed the idea of the biotic pump, have concluded that forests not only help to protect some otherwise vulnerable areas from the worst effects of powerful storms, but by softening the force of wind-borne energy they might even help suppress the likelihood of them forming in the first place. After all, if forests are capable of evaporating comparable quantities of water vapor to that coming from the surface of an ocean, and which over water can lead to the formation of hurricanes, then it might be expected that hurricane-like storms could form over forests too. The fact that generally they don't raises the interesting question as to why. Makarieva and Gorshkov, and others, believe that the explanation is down to how the biotic pump effect that is created by large areas of forests drives wind patterns that suppress the formation of storms and tornadoes.

So the clearance and degradation of the tropical rainforests is not only contributing to climate change but also taking away one of the best natural mechanisms we have for removing carbon dioxide from the atmosphere. At the same time, one of the ways that people can be protected from extreme weather is being removed and a possible mechanism for storm suppression is being degraded. All bad news—and not only for the tropical rainforests, but also what is in some ways their marine equivalent: tropical coral reefs.

Coral catastrophe: how deforestation impacts on Earth's most diverse marine systems

Tropical coral reefs are the most diverse marine ecosystems on Earth. On less than one percent of the seabed they host about a quarter of all marine species. Warm water corals are found in shallow seas across the tropics and support a vast range of life forms including many kinds of fish, dolphins and turtles. Built by sedentary little animals working in cooperation with microscopic algae, in some places their carbonate structures can combine to form huge features, such as the Great Barrier Reef that runs for 2,300 kilometers along the eastern side of Australia, or that in the Caribbean which runs parallel with the coast of Belize.

Tropical coral reefs are imperiled by deforestation and climate change
(Terry Hughes/WikiCommons)

Shores bordered by intact rainforests generally have pure clear water, but in those areas where they've been cleared or logged there is more soil erosion and thus more water-borne sediments. When these microscopic particles get into the sea they can be a factor that causes the death of the corals and the collapse of entire reef communities, one reason for which is that sediments rich in organic materials can render corals more susceptible to bacterial infections. The death of corals can lead to coastal areas becoming more vulnerable because, like the mangrove forests that once fringed many tropical coasts, healthy coral reefs can help mitigate the impact of extreme weather hitting the land. Indeed, coral reefs are in many places vital for the protection of coasts from threats arising in the sea, such as the devastating surges of water that sometimes accompany storms.

A range of stresses are lined up against the coral reef but some scientists believe that it is the direct consequences of deforestation that, at the global level, pose the greatest overall threat because of the loads of sediments that are released as fires, plows and chainsaws lay bare once-forested land. Take the case of Indonesia, which is at the center of an area of warm seas that hosts the so-called Coral Triangle, the greatest extent of coral on Earth. Around its coasts, dotted with more than 17,000 islands, is nearly one fifth of the world's coral reef—over 50,000 square kilometers of it. These reefs are not just extensive, but extraordinarily diverse—and many are still unmapped, both physically and biologically.

The marine diversity across this vast region is under mounting pressure, however. The World Resources Institute has estimated that some 85 percent of the coral reefs within the triangle of sea from the Solomon Islands to the Philippines are threatened. The forests and corals are going down hand-in-hand and in the process leaving communities more vulnerable. And, as the deforestation continues, less direct but nonetheless equally profound consequences for the coral reefs are unfolding.

More than 90 percent of the heat being trapped by elevated greenhouse gas levels is ending up in the oceans, and so is about a third of the carbon dioxide we release. When that gas gets absorbed into the water it forms carbonic acid. The more carbon dioxide we release, the more it gets into the oceans and the more acidic the waters become, as well as warming up as heat from the atmosphere is absorbed. The acidification is a problem for a wide range of sea creatures, especially those that build carbonate shells and skeletons—oysters, clams, urchins and various plankton—in turn causing knock-on effects for creatures which eat them, such as fish. Corals are affected by acidification and also high temperature, with the damage causing changes to entire food webs and ecosystems.

Both pressures cause the corals to become stressed to the point where they become 'bleached.' This is a term applied to corals where the symbiotic relationship between the microscopic photosynthetic algae that live with the little polyp animals that build coral reefs has broken down, turning the corals a ghostly white, hence the use of 'bleaching' to describe what has happened. I've seen this in different parts of the world and listened to the alarm of coral reef experts.

Even though the forest bordering many coral reefs is still in good shape, and the corals there not so far stressed by sediments entering the water, they are nonetheless suffering partially from the effects of deforestation, due to the elevated atmospheric carbon dioxide concentration. The rate of ocean acidification is the most rapid seen for many millions of years, perhaps since about 250 million years ago, when the biggest mass extinction of life ever to have occurred on this planet took place.

The effects of warmer and more acid seas are why coral bleaching events have recently hit reefs right around the world, even in places very remote from other human influences—in the Caribbean, the Indian Ocean and in the Pacific Ocean. High ocean temperatures during 2015 and 2016 killed large areas of the Great Barrier Reef, with some sectors having more than half of the corals wiped out. It was the latest in a series of severe bleaching events, following those that occurred during the summers of 1998, 2002 and 2006. The death of the corals occurs in proportion to the severity of the heat stress they suffer and, while they can recover, that's only possible if they don't get hit again, and again, and again.

All this provides a powerful reminder that it is not only to avoid direct damage to such sensitive marine ecology that we need to conserve the rainforests, but also to maintain the entire Earth system and to better appreciate how degrading one ecosystem can cause impacts elsewhere. Perhaps we need to adjust our collective perception of how all this works, and maybe to revisit some of the ideas that shape our reactions. For example, much of the debate about climate change is concerned with fossil fuels, parts per million of pollution and technology choices. This sometimes conveys an impression that the Earth's climate system is a dead entity, a mechanism that can be tinkered with by humans via technology. This is, however, completely false, for the atmosphere is a living entity, one that is the product of life and fundamentally shaped by ecosystems such as forests.

This is seen in how the system, like an individual organism, has the capacity to bring itself back toward stability when stressed. Forests grow more quickly when there is more carbon dioxide in the air and react to drought by increasing the biotic pump effect. In so doing, they mitigate extremes, holding down temperature and maintaining the flows of water vapor that moisten continental interiors, all at the same time as protecting other elements of the biosphere, that intricate web within which we humans are inextricably embedded.

That ice archive at the British Antarctic Survey underlines the extent to which our choices can make a difference to that, revealing how our actions have pushed the system beyond any comparable state seen on our planet for a very long time. As the consequences unfold, so we see again and again how causing damage to one set of systems can have major knock-on implications for others.

As we look to the future and how to maintain the carbon, climate and water services provided by the tropical rainforests, we must adopt approaches that reflect the fact that these essential natural processes are not simply the result of chemistry and physics, but also extraordinarily intricate ecology.

3 THE ECOLOGICAL WEAVE

Tropical rainforests are complex ecological tapestries that work through a multitude of intricate connections

The tropical rainforests are more than simply collections of trees but intricate systems working on the basis of many connections. Some are microscopic in scale, others planetary. When it comes to some of the global connections that shape their character, it is possible to glimpse these very far away from the forests themselves.

I saw evidence of one such connection, at home in Cambridge, on a bright morning in April 2014. Spring had arrived reassuringly on time with temperatures boosted by southerly winds, though the weather forecast had warned anyone suffering from respiratory problems not to exercise outside. Not too concerned, I set off by bike for meetings in town and as I pedaled along I noticed all the parked cars were covered in fine red dust. This was the reason for the weather warning. A huge quantity of it had arrived across Britain, blown on winds from North Africa and the Sahara desert. Whipped up by strong winds funneled between mountains, tiny particles of loose material had been taken aloft, conveyed thousands of kilometers north on a high-altitude airstream, and deposited in overnight rains.

These dust-laden airstreams are rare in Britain as they tend to head west rather than north, out over the Atlantic Ocean. They more often pass over in the Canary Islands, where they bring conditions known as *calima*—'haze'—when the dust is sometimes so thick as to drastically reduce visibility. Much of it originates from a huge low-lying area in Chad called the Bodélé Depression, the world's biggest single source of dust. At the southern edge of the Sahara desert, 2000 kilometers from the coast of West Africa, this was once a vast lake and at its greatest extent, during an especially wet period after the last Ice Age, it was the size of California. The existing Lake Chad is a tiny remnant (and one that is still shrinking).

The dust that leaves the Bodélé Depression on the wind is made from dried-out sediments—the remains of microscopic diatoms that once thrived in that huge freshwater body— that accumulated at the bottom of the ancient lake. The material from this most desolate place is rich in phosphorus and iron and other nutrients vital for the manufacture of proteins and plant growth. It travels high in the atmosphere and falls among other places across the Amazon basin, where the nutrients carried within it fertilize the forest. Research by NASA based on satellite measurements reveals that each year nearly 30 million tonnes of Saharan dust arrives there. It includes about 22,000 tonnes of phosphorus, which is estimated to be more or less equivalent to that removed each year by rainwater run-off.

The wind-borne fertilization of the South American rainforests presents one example of the intricacies of the linkages that enable our planet to function as it does: the largest desert on Earth connected to the largest rainforest—the wettest and driest joined—and in so doing affecting the functioning of the entire planetary system.

Dust from the Sahara Desert passes over the Canary Islands (NASA)

The dynamic nutrient knife-edge

The fact that the input of airborne nutrients makes so significant a difference to rainforests, replacing those lost by leaching of rainwater, underlines how rainforests are often balanced on a nutrient knife-edge. Many rainforests grow on the ancient stable soils of continental interiors that have been free from geological disturbance for tens of millions of years. As a result the ground beneath them has undergone a progressive depletion of mineral nutrients. For example, soils that underlie much of the Amazon basin rainforests have been weathered to the point where phosphorus, potassium, calcium and magnesium, all ultimately derived from geological sources, are so heavily depleted that the land beneath the trees is impoverished.

This can seem hard to believe given how the verdant rainforest vegetation gives an impression of high fertility, a notion that has often fooled colonists into clearing land for farming, only to find that after a short time yields drastically decline. High crop output might be sustained for a few years but when the soils become exhausted crops can no longer be produced. What those colonist farmers don't realize is how in rainforests the nutrients are mainly held in the trees and other vegetation above the ground, rather

Fungi in a Costa Rican forest. The warm moist conditions permit the rapid recycling of organic material (TJ)

than in the soils beneath. Typically, some 75 percent of rainforest nutrients are held in the plants, about 17 percent decomposing in the leaf litter on the forest floor and as little as 8 percent in the soil. It is very distinct from temperate zone woodland which has a far bigger proportion of nutrients in the soil and also marked seasons, with long cool periods, that mean it takes longer for decaying organic matter to be recycled and made available to plants via their roots.

Although most tropical rainforest trees are broadleaved species the majority are not deciduous in the sense of losing all their leaves at once. Most of the myriad tree species replace worn-out leaves by dropping a few at a time. When those leaves and other dead plant and animal material reach the forest floor, so the warmth and humidity intensify the rate of decomposition, and nutrient recycling, with much of the breakdown of organic material taking place above ground in the leaf litter before plant material makes it into the soil proper. Rainforest trees and plants need to keep their roots close to the surface, going out in a broad flat disc rather than downwards, which explains the need for the huge buttress structures at the base of many tall trees, enabling shallow-rooted and poorly anchored giants to remain stable, especially those that brave the winds above the canopy.

Visitors to the rainforest often remark on the stillness and calm beneath the forest canopy. But this is an illusion, for, despite the impression of stability, change is constant. Walking in a Central American rainforest one day I heard the most almighty crash. A short way from where I stood a huge branch from a canopy tree had snapped off and fallen 40 meters to the forest floor. It seemed the soaking of the bromeliads, mosses, ferns and everything else that clothed that branch had, during the previous evening's rain, taken on more weight than its 10 tonnes of rotting bough could bear.

It was a reminder of how tropical rainforests are dynamic environments. Open patches created where branches and trees fall allow light to reach the forest floor. When this happens plants seize the opportunity to grow, and to reach the life-sustaining sunshine before others do, for, in addition to the struggle to secure nutrients, a key factor for success is access to light. Seeds that lie dormant, waiting for stimulation from the Sun, burst into life. Fast-growing species dominate, creating a dense tangle of plants at ground level. At the same time as the race for the light there is a rush to feast on the wood and other plant materials that fall to earth. A log that had crashed to the ground some time before was covered with hundreds of delicate little white mushrooms, arranged like intricate Victorian porcelain.

One person who knows about the dynamics of these ecosystems is the celebrated rainforest expert and historian John Hemming. Now in his eighties, he has accumulated the kind of rare insight that comes from more than six decades of study, including (as Director of the Royal Geographic Society) leading one of the biggest research projects ever undertaken in a rainforest—the Maraca project in the Brazilian Amazon. I've known John for years and met him at his London home, decorated with a wealth of artifacts from a lifetime of travels and study in tropical rainforests, to discuss his thoughts. There were indigenous headdresses resplendent with the feathers of toucans, parrots, egrets and eagles, Victorian-era prints depicting waterfalls, rainforest wildlife, and battles between Indians and European explorers. He told me about a more recent conflict that took place in 1961, when he and his friend Richard Mason were part of a joint British–Brazilian expedition into an unexplored part of the Amazon in central Brazil, near to the Iriri river. Having been assured that there were no tribal people in the area being

Rainforests are powered by sunlight. Kakamega rainforest, Kenya (TM)

surveyed, Mason was ambushed and killed by Paraná Indians on the trail they'd cut through the forest. Hemming told me that the Paraná were unusually belligerent: 'I can think of no other people who would have lain an ambush and killed before any form of contact.'

Aside from his exceptional first-hand knowledge of the peoples of the Amazon, Hemming has built up a very clear picture of how the rainforests work ecologically. 'Rainforests run on two cycles,' he explained. 'Nutrient cycles, which is dead leaves and dead monkeys and everything else decomposing, and then alongside that a hydrological cycle—rain, water dripping off leaves and the rivers and so on.' He warned that 'if you strip away the cover any kind of topsoil goes very quickly, as a result of heavy rain,' leaving little more than what he called a 'pink parking lot.' He explained why this was such a major issue, for the soil is more than simply an accumulation of weathered rock and decomposing organic matter, but also a complex ecosystem in its own right.

During the in-depth research work that Hemming had led at Maraca he'd become aware not only of the vulnerability of rainforest soils, but also their complexity and how it is the many living elements within them enable them to function as they do. This included a growing appreciation of the roles played by threads of fungi that live in the soil, 'the mycorrhyzae that are transporting the nutrients once they've been released, transferring them to the roots of the trees.'

Mycorrhizae: the wood-wide web

The decomposition and recycling of nutrients from decaying wood and leaves is undertaken by a wide range of organisms, including invertebrates such as millipedes and termites and microorganisms in the form of bacteria and fungi. Nutrient-rich animal droppings are reprocessed with the help of dung beetles. Once nutrients are released, the ability of plants to get hold of them is in part determined by their relationship with soil organisms, especially those intricate networks of fungal threads.

These mycorrhizal filaments, which closely entwine with tiny rootlets and root hairs, form partnerships with the trees and plants that—with their green leaves held aloft on trunks, stems, branches and twigs—make food from sunshine. The fungi are supplied with some of the sugar manufactured from photosynthesis, and in return transfer nutrients to the plants' roots. These fungal root extensions do their work by releasing powerful enzymes that help to split off tightly bound soil nutrients so that plants can more easily get hold of them. This relationship is especially important, as the fungi facilitate the plants' uptake of phosphorus, a nutrient that is very scarce in many tropical soils and without which growth becomes limited. The fungal threads are also highways along which bacteria and other microorganisms travel through the soil, helping to enable them to play their parts in nutrient recycling.

What is perhaps even more remarkable is how the threads also connect plants with one another, providing a sub-soil communications network—a phenomenon some biologists have termed 'the wood-wide web.' Through mechanisms that are yet to be fully explained this underground connectivity has a major bearing on the overall mix of plant species found in particular areas of forest. There is also evidence suggesting that mycorrhizal connections help plants fight off pest attacks, with warnings of infestations passed via chemical messages from one plant to another, thereby permitting molecular defenses to be raised in advance of a full-scale assault. This communication appears to occur not only among plants of the same species, but between different ones. Perhaps the fungal threads facilitate this kind of communication because the mycorrhizae have an interest in keeping the whole forest healthy, for if the conversion of sunlight to food is diminished by pest damage, so are the fungi's prospects.

Similarly fundamental and intricate ecological relationships vital for the functioning of the rainforest go on above ground.

An ecological tapestry of connections

The tendency of the Western mind is to seek understanding of complex systems via their separate components, to categorize the individual species, to see each part in isolation. In the rainforest we are attracted to the huge trees, butterflies, birds and other eye-catching expressions of life. We are drawn to name and categorize these separate elements in the system, with the inevitable consequence of forming a fragmented view of what it is and how it works. The more we look, however, and the more one absorbs the reality of the forest, the clearer it becomes that it is far more than a collection of individual parts, but rather a tapestry woven from threads of sunshine, water, air, nutrients, photosynthesis, color, sound and genes, all united into the fabric of an inseparable whole.

Take the attractive flowers that adorn so many rainforest trees, shrubs and herbaceous plants. They are essential for plant reproduction through the transfer of pollen and growth of seeds and fruits. Over the past 160 million years or so plants have developed ever more sophisticated flowering strategies to ensure sexual reproduction. Mixing up genes creates variety, and that conveys huge advantages for survival. Should conditions change, or a new niche become available, then with variety it is more likely that one or more offspring will be able to exploit the opportunity, or adapt to changed circumstances. This is why plants, like animals, go to great lengths to reproduce sexually, to fertilize seeds with pollen from another individual.

One way to move microscopic pollen capsules (the plant equivalent of sperm) is by casting them to the winds. Producing a lot of it and hoping for the best is a viable strategy for some plants. Grasses are wind-pollinated and for these it is a winning way to go, not least because in many species individuals grow close to one another in open habitats where air can move pollen grains unimpeded. Grasses aside, however, the vast majority of advanced plants use animals to boost the success of pollen transfer. This is especially the case in tropical rainforests, where beneath the canopy there is little wind and a high diversity of species, with individuals of the same kind often widely dispersed from one another. Both factors limit the effectiveness of indiscriminate pollen broadcast, as do high humidity and frequent rains.

It is insects that are the key pollinators, but not only bees. Wasps, beetles, butterflies, moths and others are connected to plants via this route too. The huge diversity of insects in tropical rainforests says a lot about the intricacies that have evolved between the life cycles of so many animal and plant species. That interdependence comes down to the plants' interest in pollen transfer being matched by a parallel interest among animals for eating sugary nectar and protein-rich pollen.

Although animals and plants have overlapping interests, their interests are nonetheless different. Animals want easy sources of food, while the plants are better off making as little of such energy-intensive resources as they can get away with. Animals want a big meal from each flower, whereas the plants want the animals to move on and to visit as many flowers as possible. Animals benefit from big showy flowers that advertise where to go, whereas plants want to expend minimum effort on making them. The colorful interactions that take place in the forests are an accommodation to these and other ecological balancing acts.

For example, while the plants welcome the job of pollination being done by animals, they don't like their leaves being eaten. The fact that some insects do both at once makes for an even more complicated juxtaposition of interests. Butterflies and moths, for example, play vital roles for many plants, flying between blossoms in search of nectar, but their caterpillars are less welcome, as they munch through foliage. Therefore plants grow flowers to attract the adult insects as pollinators, while at the same time manufacturing toxic chemicals to deter their

Black-cheeked woodpeckers occupy an extensive range from southeastern Mexico to Ecuador. In addition to insects they also eat fruit and nectar, playing roles in pollination and seed dispersal (TM)

larvae. Some of these larvae accumulate plant-manufactured toxins to render themselves poisonous to predators.

The many interactions going on between animals and flowering plants in relation to pollination are not confined to insects, of course. More than 500 species of plants are known to rely on bats to pollinate their flowers, including some of the most valuable exports from tropical countries—bananas, mangoes, cocoa and agave (the source of tequila). Bats are also important pollinators of commercially important timber species.

Plants that rely on bats for shifting their pollen around have developed particular strategies to maximize their success. For example, the flowers of *Oroxylum* trees produce nectar only in tiny bursts, so as to encourage bats to visit a larger number of flowers. Species that attract bats also have flowers that open at night, have a strong scent, and produce viscous nectar that can be easily lapped up by a mammalian tongue.

Many tropical rainforest plants are pollinated by birds, with hummingbirds the best-known group. They seek out nectar from flowers by day, and so are using their eyes, requiring that the flowers they pollinate are bright and produce watery nectar that can be sucked through their long fine bills. Some plants have evolved strategies to make the most of these birds' attentions. For example, some *Heliconia* species that share the same bird pollinator produce flowers at different times of the year. This reduces competition for pollination services as well as the risk of unwanted hybridization. Some of the honeycreeper birds that live in the rainforests of Hawaii are also nectar feeders and have, over millions of years, evolved close relationships with native trees, from the flowers of which they take nectar.

Whether achieved by insects, birds or mammals, once pollination has taken place plants produce the seeds needed that will found their next generation. These sources of new plant life are often embedded inside fruits, many of which are also explicitly designed to attract animal attention.

Dispersal: how and why seeds move around the forest

A wide range of rainforest animals move seeds around. For many kinds of trees, shrubs and herbaceous plants this interaction is a key element for their on-going survival, including many of the massive emergent trees that tower over rainforest canopies. Consider fruit-eating birds such as hornbills, which consume a wide variety of food items and which via their digestive systems spread seeds far and wide throughout the forest. Without their broad wings and taste for fruit, many forest trees would not be able to reproduce. Sulawesi red-knobbed hornbills, for example, feed their young on the fruits of at least 33 species of plant, carrying food to the nest from up to a kilometer away. Discarded, regurgitated and defecated seeds germinate near to the nest while others are spread more widely across the forests by the foraging birds.

Big mammals—pigs, elephants, rhinos and tapirs—are also important seed dispersers for large-sized tree seeds that need to pass through the gut of an animal before they'll germinate. Elephants have been described as 'mega-gardeners,' due to the way they move tree seeds through the forests, not only taking them long distances but also depositing them in a generous package of manure. African rainforest elephants are especially important ecosystem engineers, with many of the species of tree whose seeds they disperse having evolved together with the animals. In the Tai Forest National Park in Ivory Coast, forest elephants are the major (and perhaps even the sole) dispersers of seeds from up to thirty kinds of trees.

Many other animals help with long-distance seed movements— bats, toucans and fruit pigeons among them—and some very

A collared aracari feeding in a fruiting tree at La Selva, Costa Rica (TJ)

long-distance seed movements depend on migratory birds. Even fish that live in rainforest rivers help with seed dispersal. On the island of Borneo alone, thirteen species of fish have been found to feed on fruits and to be important seed dispersers. Giant fish living in the seasonally flooded forests that flank the mighty Amazon river are also fruit eaters.

Many seeds not only require animals for transport but also to trigger germination. Gastric fluids soften the tough casings that protect some seeds against them being eaten by parrots and other seed predators before they are ready to fall from the tree. The process of passing through the gut of a mammal or bird is thus, for many trees, not only a vital aspect in breaking seed dormancy and kick-starting growth but also reducing the risk of seeds being destroyed.

In a primary forest in the Caribbean lowlands of Costa Rica I came across one of those immense straight-stemmed trees and found that its hard fruits had dropped to the side of the boulder-strewn river. They were about 7 centimeters across and with a hard almost impenetrable shell that needed some rough treatment from a machete to reveal the pale-grey (and very tasty) gelatinous fruit inside.

Amazingly, seed- and fruit-eating pacas and agoutis have the dental equipment needed to penetrate these tough capsules and the tree had created a fruity incentive for them to do so. The tree relies on these animals to eat the fruit and swallow the seeds and for these to pass intact through the gut and to be deposited in droppings at a different place in the forest. Some trees that make hard-shelled seed casings not only rely on animals to break them open and swallow the seeds whole, but also to break them open and then hide them. Agoutis do this for Brazil nut trees, cracking open the neatly packaged seeds (nuts), hiding them and later forgetting where some of them are.

Without such hard shells a lot of other animals would be able to get at the fruit, and some of these would eat the seeds, crushing them rather than hiding them or passing them through their digestion undamaged. But even with heavy-duty protection like this, some seed eaters can still get in. Great green macaws, for example, which like many other parrots are seed predators, have powerful bills designed to crack the toughest seed protection and often discard fruit in favor of the more nutritious seeds. It's one more balancing act that natural selection requires the trees to master.

Trees that produce fruits often attract a wide range of species. In one Costa Rican rainforest on the country's Caribbean lowlands

I was struck by how many birds (and in some cases wild pigs, in the form of collared peccaries) could be found feeding on a single fig tree. Tanagers, toucans, aracaris and thrushes were among the diverse range of species that congregated to feast on the nutritious fruit. Another popular species that had attracted many birds and mammals, including three-toed sloths, was the *Cecropia* tree. These could be found in many forest openings and they grow very fast to exploit the abundant light under a broken canopy after a tree or big branch fall. Their seeds need to be in position to await such an opportunity, and animals are vital in making that happen.

When they've germinated, the *Cecropia* trees also rely on animals for another vital service: protection. These trees attract *Azteca* ants through providing them with nesting cavities in their hollow stems and a fatty secretion that the insects love to eat. The ants drive off herbivores that might eat the trees' foliage, including leafcutter ants, and help prevent the establishment of epiphytes that would steal some of the trees' hard-won light. Their waste and dead bodies also contribute nitrogen fertilizer that helps the trees to grow.

Among the less likely seed dispersers that I noted were Honduran tent-making bats. I came across a group of these tiny little white bundles of fluff asleep inside a folded banana leaf. With wings tucked away and tiny pink ears visible, they looked like tail-less white mice. The animals had bitten along the long central spine of the huge leaf, causing the two halves to flop together, creating a neat hiding place and shelter. Inside, their white fur absorbed the color of sunlight coming through the leaf, making them look green, further aiding concealment. At night these little creatures forage for fruits, in the process swallowing seeds, which they spread around the forest. The dispersal process is enhanced by the bats' anti-predator behavior. This is because

Honduran tent-making bats at rest in a daytime roost (TJ)

they make several of their leaf shelters, moving from night to night so as to avoid ambush from a snake or cat.

John Hemming explained one of the reasons why the trees go to such lengths to build relationships with animals, not only to achieve sexual reproduction but also the dispersal of their seeds: 'Tropical rainforests have hundreds of tree species, each with its own strategy for seed dispersal and pollination; and the birds, monkeys, rodents and insects that perform that dispersal and pollination is each part of a different food chain. The reason every tree devotes so much energy to this is because there are so many blights, parasites and enemies attacking them, and most of these are species-specific.' By building specific relationships in such a diverse system, trees can keep pace with some of the pressures ranged against them.

Some animals that eat seeds for food can also be helpful to plants. Various kinds of rodents store seed in caches hidden around the forest. If the rat or squirrel that created the store gets eaten, or forgets where it hid it, then the seeds can germinate far from the tree that grew them. As is the case with pollination, seed dispersion is highly dependent on animals, with some 85 percent of woody rainforest species relying on fruit-eating birds and mammals to do the job for them.

Rodents and other animals are also important for the perpetuation of an even more fundamental element. The spores of the mycorrhizal fungi that are so important in nutrient recycling and distribution have been found in the feces of many rainforest animals. Some mammals—for example, kangaroos and bandicoots in the Australian rainforest—seem to accidentally ingest the spores while they are foraging for food, while others actively seek out the fungal threads to eat; in both cases they disperse the spores via their droppings.

It's not only vegetarian animals that are of importance to the functioning of the forests. The mammals and birds that eat fruits and vegetation are in turn eaten by predators, such as tigers and jaguars. These keep a check on the abundance of herbivores such as deer, pigs and tapirs, and that in turn influences tree regeneration. Ant-eating pangolins have potentially profound impacts on how forests work, given the vast number of ants they consume. Ants are vital for a range of fundamental processes ranging from pollination to decomposition and there can be little doubt that the dynamic relationship between them and the pangolins influences wider forest ecology.

These ecological relationships also connect back to how forests shape the composition of the atmosphere. For example, one group of researchers found that, by simulating the local extinction of trees that depend on large seed-dispersing mammals in Brazil's

Atlantic forests, carbon storage was significantly eroded, even when only a small proportion of the tree species were lost following the disappearance of the animals. These researchers noted that while international efforts to reduce carbon emissions from forests were focused on reducing forest loss, 'our results demonstrate that defaunation, and the loss of key ecological interactions, also poses a serious risk for the maintenance of tropical forest carbon storage.'

Many people would regard efforts to crack down on the illegal ivory and rhino horn trade, the unlawful hunting of monkeys and the conservation of toucans and hornbills as separate issues to that of seeking to tackle climate change. However, the evidence is beginning to reveal that these things are actually fundamentally connected to one another. Forests are more than simply collections of trees, but rather webs of connections within which the trees are (to our eyes, at least) the most obvious components. They are connected by mammalian nuts and bolts, the ecological glue of birds and the strings and struts of insect life. That pollen between the flowers makes fruits that are moved by the tapirs, pigs and the rest, and the huge trees in turn lock up the carbon and pump water into the atmosphere, powered by sunshine and nutrients (some of which arrives on the wind from faraway deserts).

Convergence: why wildlife in diverse rainforest can still look alike

The extent to which the different roles fulfilled by animals are integral to the overall rainforest system is underlined by how similar-looking creatures that do similar things have evolved independently in different parts of the world. Thus, while the tropical rainforests have been incubators for the increased diversity of life on Earth, they have also converged: that is to say, created tendencies that lead to living things looking similar, even though they are not closely related.

The process that leads to this phenomenon is known as convergent evolution and it is driven by how similar niches and functions exist in broadly comparable ecosystems, no matter which part of the tropics they are in. For example, the sunbirds you find in Africa and Asia are very similar in appearance to the hummingbirds of the Americas. In the forests of Central America I have been struck by the similarities between motmots and the bee-eaters that I had seen across Africa and Asia. In Indonesia, in remote areas where rainforests remain, there is nothing more memorable than the calls of gibbons at dawn. Gibbons are apes and there are no such animals in South and Central America, but large primates (including howler monkeys) have evolved to fill ape-like niches, complete with the dawn calls that carry for great distances through the forests.

Although harder to see, most tropical rainforests also have their big predators: leopards in Africa and Asia, tigers in Asia and jaguars in the Americas. All these big spotted cats are so-called keystone species, shaping the rest of the life in the forest, creating cascades of effects through the ecosystem via their special role as top carnivores.

The tropical rainforests thus embody an incredible complexity of ecological relationships, in turn sustaining water and atmospheric processes of global importance. At the same time as evolving their intricate ecological character, they have also become Earth's most diverse terrestrial ecosystems.

That diversity has been a long time in the making, and we're increasingly in a position to appreciate just how important that is, as well.

4 EVOLUTIONARY TREASURES

The most diverse ecosystems on Earth, rainforests comprise living assets of incalculable value

Equatorial rainforests have been evolving on Earth for almost 200 million years. Back in that distant past—the early part of the Jurassic era—most of the world's land was concentrated into a single supercontinent called Pangaea. As the world became warmer and wetter, so it was increasingly forested, though the forest was radically different to those of today. Flowering plants had not yet evolved, nor mammals nor birds. Instead, forest canopies comprised giant conifers, with tree ferns, palm-like cycads and ginkgo trees making up the lower storys. Reptiles and amphibians, some huge, stalked the forest floor, while winged pterosaurs flew overhead, scanning for prey.

Modern rainforests still contain reminders of those primeval jungles: tree ferns grow in many tropical rainforests, while *Araucaria* trees, ancestors of monkey puzzle trees, are still found in the forests of New Guinea. But the extreme biological diversity of today is in part the result of the movement over many millions of years of the continents. Pangaea broke up about 175 million years ago, with the landmasses of South America, North America and Africa becoming vast islands, leading to divergent evolutionary pathways and the evolution of distinct flora and fauna. When much later some of these isolated landmasses once more collided with one another, so evolutionary forces were again switched to overdrive.

An ecological superhighway

The Pan-American Highway stretches for almost 30,000 kilometers from Alaska in the north to the ice fields of Patagonia in the south. Around the border of Nicaragua and Costa Rica it crosses a relatively slim stretch of land that is part of the Isthmus of Panama. This is where an ancient, ecological superhighway was completed when, 2.8 million years ago, North and South America became connected by a land bridge.

This intercontinental connection started with a string of volcanic islands—part of the Pacific Rim's so-called 'Ring of Fire'—whose rise was followed by an uplift of the seabed, as two tectonic plates from the Earth's crust ground together. As the Caribbean Plate slid over the edge of the neighboring Cocos Plate, beneath the Pacific Ocean, a stretch of unbroken dry land linked the Americas. This precipitated an event called the Great American Biotic Interchange, when animals and plants that were previously isolated from one another were able to mix, enabling wildlife originally from South America not only to colonize the adjacent northern tropics, but to eventually reach the temperate regions of North America, and vice versa.

Plants made the crossing first, with their seeds island-hopping in both directions: floating in the sea, being carried on the wind or transferred by birds. Most animals, especially those that couldn't fly, needed a more substantial crossing point, with the majority making the connection later, when the bridge was complete. Opossums, armadillos and porcupines were among those moving north. Bears, cats, dogs, horses and raccoons made the trek southward. It was an event regarded by geologists and ecologists as one of the most profoundly important to have taken place on our planet—certainly the most significant during the last few million years. It had major implications for forest wildlife in the Americas and as far afield as Africa and Asia.

This was because the land bridge that linked two vast continents also separated the Atlantic and Pacific—in the process changing ocean circulation and creating one of the factors that led to the Pleistocene glaciations. Vast ice sheets advanced and retreated about twenty times, ending with the conclusion of the last Ice Age around 10,000 BCE. For long periods it was much colder, and drier, for, whereas warm seas often bring rain, cold ones bring drought. And because so much water was in the ice sheets that over vast areas were up to 4 kilometers thick, sea levels were much lower than they are today, at the peak of the glaciations—about 140 meters below where they are now.

Tropical rainforests survived the drier conditions, but they became fragmented, some effectively becoming archipelagos of moist islands set amid seas of drier forest and grassland. Africa's rainforests at times shrank to a fraction of their former extent, with the vast Sahara and Kalahari arid zones becoming united with one another. During the warmer interglacial periods, the wet forests formed a broad continent-wide band, spanning the equator from the Guinea coast in the west to Mount Kenya in the east. Amid all these upheavals were forests that remained wet throughout, including forested mountain ranges such as the Usambara in Tanzania.

While drier conditions led to 'islands' of rainforests being created on continents, the reverse happened to some other rainforests when actual islands became connected to continents. Each time ice formed and the sea level dropped, many plants and animals could move to places that were previously isolated by sea. This is why, for example, elephants, tigers and rhinos made it

The interdependence between different animal and plant species is one reason for increased diversity in the tropical rainforests. Female unidentified moth, Ecuador (TM)

from the Asian mainland to islands in modern Indonesia. When sea levels rose again, plants and animals that had moved to new lands once more became separated on islands, many giving rise to new species that would be found nowhere else.

Mega-diversity

After about two and a half centuries of systematic inquiry, scientists have described about 2 million different species. But we continue to find new ones all the time, especially in tropical forest ecosystems. An authoritative 2011 study estimated there to be a total of 7.4–10 million species on Earth (not including bacteria and viruses), suggesting that so far we've named between a third to a fifth of Earth's lifeforms. We can't be sure how much of this variety is found in the tropical rainforests, but the established wisdom is that it is more than half of terrestrial biodiversity—an extraordinary concentration in a set of ecosystems that in their original state occupied just 8 percent of the land.

This exceptional species richness is down to the particular combination of circumstances that prevail in tropical rainforests. As we have seen, one factor is that they are supercharged with an abundance of sunlight and freshwater—two of the essential prerequisites for complex terrestrial life. The wet and sunny conditions are made all the more conducive to growth and reproduction through warmth and the absence of marked seasonal change, including in cooler upland tropical rainforests where life flourishes all year long, increasing the opportunities for extreme specialization.

The fact that tropical rainforests have also been around for a very long time, with the mix of species that comprise them changing as environmental conditions have altered over millions of years, has also boosted diversity, shaped by the epic effects of geological isolation, connection and reconnection. In the moist tropical conditions, the complexity also fed off itself. For example, when flowering plants emerged and over time developed ever more intricate relationships with insects and other animals, so the forests became more diverse.

The diversity is seen most obviously in the many different species, but it is also reflected in the subtle differences between the different varieties within those species. Such variants, isolated by major rivers, mountains and stretches of sea, are in turn comprised of local populations, all of which can be distinct to others, even in neighboring valleys.

These are the reasons why the tropical rainforests are so diverse compared with other ecosystems that emerged more recently in Earth's history, in colder, drier or more seasonal geographies. The differences between the tropical rainforests and some other kinds of forests are huge. Whereas, for example, a temperate forest might be dominated by just a half-dozen species of tree, a tropical rainforest can have getting on for 500 in a single hectare. In Peru, more than 1,300 species of butterfly have been recorded from just one National Park—more than three times the entire number found in all of Europe.

Where that land bridge was established in modern-day Costa Rica it is possible to see just how much diversity can be packed into a relatively small area. Costa Rica occupies an area of just over 51,000 square kilometers, less than a quarter the size of Britain. Yet the variety of wildlife found is so much greater. There are, for instance, over 800 bird species (Britain has less than 300); 174 species of frogs, toads and salamanders (Britain has just 7); 137 species of snake (Britain has 3); and 110 kinds of bats (Britain has 18). When it comes to plants, Costa Rica has about 12,000 species, while Britain has fewer than 3,000.

So it is that the fundamental forces of plate tectonics and climate changes have over a vast expanse of time combined with tropical sunshine and moisture to nurture into existence the most diverse ecosystems on Earth. And the rainforests' extreme species diversity is also a consequence of the fact that they themselves are quite varied. As many as forty different kinds of lowland rainforest have been identified, with each characterized by a distinct combination of rainfall, soil and drainage.

The many species of trees that dominate lowland forests together form a canopy at about 45 meters above the forest

Stands of mangrove trees line the coast in front of a primary lowland rainforest at Piedras Blancas National Park, Costa Rica (TJ)

floor. The species that break through the main canopy—so-called emergent trees—often reach 60 meters, with straight unbranched trunks up to 40–50 meters before the branches that sustain entire ecosystems radiate out above the main treetop canopy. Some emergent trees are truly immense: a yellow meranti (*Shorea faguetian*) in the Tawau Hills National Park, on the island of Borneo, is recorded as more than 88 meters tall.

Other rainforests cover the sides of mountains. Some are in the subtropical zone, while others at higher altitudes are cooler and apparently temperate in character. In the uplands, above about

900 meters, there is generally a lower canopy and, instead of the trees having tall, straight and unbranched stems, they tend to be more gnarled and often multistemmed. Evaporation is slower and so is growth. Decomposition is slower too and nutrients are released more gradually. There are no termites and fewer ants and there is thus a bigger role in the decomposition process for earthworms and beetle larvae.

In many cool upland tropical rainforests biological production exceeds the rate of decomposition, and in some moist valleys deep layers of saturated peat have accumulated, holding not only a great deal of carbon but also vast quantities of water, both of which shape its soil and indeed the character of the rivers that flow from them.

So, while the term 'tropical rainforest' is used to describe a set of ecosystems that share some broad common characteristics, they are very varied, even across relatively restricted geographical areas. Take the forests that lie today across that ecological superhighway in Costa Rica . . .

Shore to summit: how rainforest changes with altitude and aspect

The Rio Esquinas flows into the sea on the Pacific coast of Costa Rica, where it forms the northern boundary of the Piedras Blancas National Park. On the flat lands that flank its lower reaches are extensive stands of mangrove forest—comprising seven species of the marine trees that give the ecosystem its name. The architecture of these special trees enables them to thrive in low-energy marine tidal environments. Saltwater forests like these are found between the sea and rainforests on land right around the tropics.

The mangroves' arched multiple stems allow water to pass between the trees while trapping the silt in which they grow.

The entangled roots and stems create conditions that sustain not only this unique marine forest but also numerous ocean fish that lay their eggs among the sheltered nooks. Sharks and snappers come in from the deep waters of the Golfo Dulce to spawn, attracting in turn a whole host of predators, including tropical river otters and crocodiles.

As I paddled between the tangled stems, a flock of about eighty white ibis took to the air, with black wingtips carrying aloft their snowy bodies; the birds' extended necks and legs revealed their evolutionary affinity with the storks. On the pale trunk of a mangrove tree was a line of roosting proboscis bats, nine of them. As the tide began to drop in the smaller side-channels, crabs ran over the stems, emerging from burrows that were submerged at high water.

The haunting cry of a whimbrel carried between the mangroves. It was a sound familiar to me from the northern tundra and a surprise to hear amid the steamy sea forest. But it was not alone. Other travelers included spotted sandpipers and willets, bobbing about on the mud, searching for the invertebrate food that would in a month or so power their epic spring journeys back to the rich long days on their temperate and Arctic breeding grounds.

Over hundreds of meters the tangled sea forests gradually give way to forests on permanently dry land. On more exposed parts of the shore the transition between the marine environment and the tropical rainforests was almost immediate, just a few meters of sun-baked rocky shoreline marking the boundary between the world of turtles, dolphins and corals and that of trees, parrots and butterflies. I went ashore as the brutal heat of the tropical sun intensified towards midday, but upon entering the forest I became the beneficiary of a natural air-conditioning system, cooled by the evaporation of the plants and the shade cast by the canopy of foliage held aloft on the huge trees.

Primary forest grows down to a rocky shoreline at Piedras Blancas National Park, Costa Rica (TJ)

'Primary forest' is the term given to largely undisturbed formations that have not, in recent times at least, been significantly changed by human activity. Most such forests have been influenced by the activities of native peoples, but the structure of the forest and the mix of wildlife species living in it have remained largely intact. The rainforest of the Piedras Blancas National Park is one such example, and it was full of life.

In the gloom beneath the thick shade I could see the buttress roots of two huge trees that had apparently fused together.

Straining my neck to look up it was hard to know whether the trees were of the same species, or different ones that in the exuberance of growth had made a joint attempt to reach sunlight. One reason why identification was difficult was because of the vast number of mosses, bromeliads, orchids and other plants growing on the branches of the trees. There were even small trees growing on the branches of the big trees. An entire ecosystem was held aloft. There was a lot going on at ground level too. Along one of those fused roots a small lizard chased a spider that was about three times bigger than its head. Despite its formidable bulk, the spider was taking no chances and ran up one of the trees.

A deep buzzing noise close by seemed to warn of potential hazard, but I turned my head to see a rufous-tailed hummingbird. Zipping and hovering between flowers it was one of more than a dozen species of such birds found there. It landed next to the red blooms of a flowering shrub, pausing momentarily before it sallied from its perch to barge into a yellow butterfly that was considerably bigger than the tiny bird. It seemed a deliberate act of aggression, perhaps because the insect was a competitor for nectar.

The frantic worried chattering of a different bird came from deep cover. It was hard to see, but after a while I secured a good enough view to identify a chestnut-backed antbird. There were lots of ants in the forest, and as I spent several days trying to put names to the birds living there I was struck by how many of them had the word 'ant' in their name, including another I found that day called a black-hooded antshrike: a regional endemic, found nowhere else—perhaps another product of the evolutionary stimulus created by the land bridge.

There were bigger birds too, including great curassows, the size of a turkey and with their impressive crests among the most primitive of the world's feathered beings. One walked silently into a rare patch of light that reached the forest floor. When it realized it had been spotted, it stopped and, as though trained by a fire safety expert, exited calmly into the gloom.

Even though sound is more important than sight for communication among nocturnal animals, light remains important even when it's dark. Shortly after an early evening deluge and as the hypnotic sounds of night ramped up, I thought my eyes were playing tricks. Little white lights popped on and off and would drift for a second or so in the middle branches of some of the trees. After a while I realized it was fireflies beginning their night-time pyrotechnics. They were not the only creatures in the forest using biology to create light. A clunky beetle bustled across the forest floor, sporting two bright green shining spots that looked like headlights.

Venturing between the trees with a torch revealed some of the sources of the rich array of voices: frogs, crickets and nocturnal birds among them. A torchlit encounter with a Mexican mouse opossum followed by another with a troupe of crab raccoons provided a reminder of the history of this place, a mixing of mammals, the former descended from Gondwanaland's marsupials and the other from the more advanced animals that came south from the once separate North American continent.

Upland rainforest

Heading away from the coast and up into the spine of mountains that dominates the center of Costa Rica it becomes clear that the term 'rainforest' covers a wide range of ecosystems. The tropical sun is still fierce, but in the shade of the trees it feels quite temperate, and in the early morning decidedly cold. In this tropical latitude intense productivity among plants means that dense forests grow higher than 3,000 meters above sea level, beyond which point the trees become stunted, beginning the transition into high-level shrub lands known as páramo.

The difference between the lowland forests by the ocean and the cooler temperate mountaintop habitats reflects another of Humboldt's observations, and indeed a principle named after him—Humboldt's Law. This holds that for each 90-meter increase in altitude on a tropical mountain there is an annual average temperature change equivalent to traveling about 108 kilometers north in latitude. According to this broad rule, the altitude of the equatorial páramo would be about equivalent to sea level at the latitude of Chicago, or northern Spain.

The wildlife in these upland forests confirms, however, that it is decidedly tropical. Tapirs, pumas and ocelots live there, as do many species of tropical rainforest birds, including upland specialists. In the San Gerardo Valley, high in the Talamanca mountains at 2,200 meters above sea level, there are forests very different to those of the Pacific coast. At this altitude they don't only look different—they feel different. Plodding along, slowly climbing the forest trails, the price of relief from the profuse sweating of the lowland forest is breathlessness.

Huge oak trees of three species dominate, descendants of those moving south from temperate North America during successive glaciations. They are evergreens and can grow to more than 40 meters, soaring upward on thick straight stems. Most were covered with mosses, orchids and shrubby and woody species adapted to growing on their branches. From a distance large red bromeliads growing on the oaks looked like giant flowers. I found a lot of lichens too, some hanging in fronds from branches, more growing in furry coverings on twigs.

There were many kinds of plants growing at the ground layer, but unlike the warm humid lowland rainforests I saw no palms. In some sunny open areas where a landslide or tree fall has opened the forest floor to light, there were grassy areas with flowering herbs. These little patches of sunlit ground reminded me of temperate meadows, producing scenes that wouldn't look out of place in England, or New England.

In these special forests dwelt a bird that people travel from all over the world to see. I went looking for it myself and, just after dawn, as the first grey light gathered, heard a gentle plaintive call coming from the forest—it might be described as a slow chop-chew, chop-chew, chop-chew. Through a chink in a dense web of oak branches I got a sight of the source of the call: a resplendent quetzal. With a body the size of a large pigeon, bright blue-green with a meter-long tail, this very special species of trogon is unmistakable. The quetzal was perched in a wild avocado tree, the fruits of which comprised the main part of its diet.

But while the quetzal is one of the more dramatic-looking birds found in these mountain forests, it is just one of the region's unique species. Of the nearly 900 species of birds that breed in Central America, about a quarter are restricted to the highlands and a tenth are Central American endemics—that is, found in these land bridge mountains and nowhere else. Many of them are confined to the cool forests that span the highlands straddling the border of Costa Rica and Panama.

Long slow walks revealed a number of them that had evolved up there in the isolation caused by tectonic and climatic upheavals. They include the black-faced solitaire, with its call that sounds rather like a squeaky door; tufted flycatchers, perched on twigs from which they launched sorties to catch flying insects; black-cheeked warblers that hopped around on the forest floor in search of insects; and ruddy tree-runners that scuttle along branches in search of food. In the litter on the forest floor, large-footed finches tossed leaves over their shoulders in search of invertebrates. Most impressive to my

A resplendent quetzal – Costa Rica's iconic bird – which can be seen in Central American highland forests (TM)

eyes was a group of long-tailed silky flycatchers. First located by the pings and trills of contact calls between flock members, the beautiful yellow-crested creatures fed on little black fruits, confirming that (despite their name) they have an omnivorous diet, and evidently a role in dispersing seeds.

Land of the frogs

Less than 200 kilometers away, on the far side of the Continental divide, in that Costa Rican cloud forest I described earlier, the ecology of the forest was different again. The ground was strewn with boughs or even trees that had become too heavy to bear the weight of all the epiphytes and the water taken from the clouds. The nutrients that sustain this explosion of life were being rapidly extracted from the dead vegetation and fed back into the endless and rapid exchange between life, death, decay and life, powered by warmth and moisture.

The forest was full of life by day but revealed many of its most fascinating occupants at night. As darkness gathered, the sound of the birds was replaced by the whistles, croaks, gratings, chirps and tweets emitted by a variety of insects and amphibians. With a torch and a keen ear I found masked tree frogs sitting in bromeliads. On a leaf was a dwarf rain frog and on another a common rain frog. These tiny amphibians are among a group of about thirty species that have achieved the remarkable feat of doing away with the tadpole stage of their life cycle, their eggs instead hatching directly from the adults into froglets. This reproductive strategy avoids the perils of predation that afflict the larval stages of such animals, denying fish and birds a meal and thereby requiring the adult frogs to lay fewer eggs.

Other amphibians I encountered included the slender-fingered frog (so small I could hardly see the frog, never mind the size of its fingers). Nearby was a red-eyed stream frog and then on a leaf an emerald-glass frog. Visible through its translucent body and limbs were parts of its green skeleton (this animal is credited with inspiring Kermit, the Muppet character). I discovered that a recent arrival up there was the animal whose image was printed on so many Friends of the Earth rainforest campaign leaflets—red-eyed tree frogs. They were newcomers in that forest, though, having moved there during recent years from lower elevations, possibly in response to rising temperatures.

The predators of frogs and other smaller creatures were out, too. A red-ringed snail-eater slid up a moss-covered creeper. This delicate slender snake was on the hunt for small creatures, including, naturally, snails. Millipedes were also active at night. These myriapods comprise a vital battalion in the army of recyclers who get to work on fallen branches, leaves and twigs, munching them up before microbes decompose organic material, releasing nutrients that are in turn returned to plants via soil fungi. They are thought to be among the oldest of terrestrial animals, dating back hundreds of millions of years.

Ocelots have also been recorded in that Costa Rican cloud forest. These beautiful cats spend a lot of their time in the canopy, quietly concealed among the mosses and bromeliads, where in the dappled shade their exquisitely spotted fur makes them very hard to see. I didn't see one but knew they were there because I was shown images taken from camera traps—devices that automatically capture pictures of such secretive creatures. Other hunters rely less on concealment and more on bright colors. One was a bright-yellow pit viper sitting next to similar-colored flowers awaiting hummingbirds.

In a road trip of just a few hundred kilometers it is thus possible to see several kinds of tropical rainforest—from coastal mangroves to lowland primary forest to upland oak forest to steamy cloud forest, all packed with their own characteristic

The red-eyed tree frog is a recent arrival in the Los Angeles cloud forest, perhaps moving to higher land in response to rising temperature (TM)

and unique wildlife, and all within close proximity of one another. The long history of these different kinds of forests, the upheavals that have changed them and the myriad of ecological interactions going on within them, have rendered the tropical rainforests the most species-rich terrestrial ecosystems on Earth. And, as we have seen, this exceptional diversity does not simply live in the tropical rainforest—it *is* the tropical rainforest. As is the case in relation to water and carbon, that diversity is also of huge importance to humankind and has the potential to help us solve many present and future challenges.

Innovation and biomimicry

One afternoon, whilst walking along a partially dry riverbed in a Colombian rainforest near to the Caribbean coast, I noticed a large brown caterpillar on a shrub. It was the biggest I'd ever seen and resembled a piece of bark—an impressive creature, though not as impressive as its parents. Along that riverside, gliding and flapping between the trees were the huge adult butterflies into which that fat caterpillar would soon transform.

One crossed a clearing and landed on a branch where, at rest with its wings closed, perfect camouflage caused it to disappear. When it once more took to the air to resume its lazy zigzag flight in search of the rotten fruits upon which it fed, it morphed into another creature again, embarking on what can only be described as a flash dance, as sunshine collided with the surfaces of its upper wings and was then emitted as iridescent blue flashes so vivid as to appear electrical. The butterfly was a blue morpho, a species that has become a source of technological inspiration, for those blue flashes are produced not by pigment but millions of nanoscale plates, the shape and arrangement of which causes light reflected from them to be disrupted into brilliant electric blue. Creating the same color from pigment would take more energy than the solution the butterfly has found, thereby enabling the creatures to devote more effort to other tasks, such as producing caterpillars.

The morpho's innovation has spawned 'biomimetic' research into fade-proof paints, potentially leading to products that not only last longer but which can be manufactured with less energy and fewer toxic by-products. It has also led to new research in the design of touch screens for computers and smartphones. Using structures inspired by the tiny scales on the butterfly's wing it is possible to produce a full-color display through the disruption of reflected light, meaning that the screen can run on less than a tenth of the power needed for a conventional LCD screen.

Andrew Parker is a scientist based at the University of Oxford who has devoted many years of study to how innovations created by natural selection can assist in technology. 'Biomimetic innovations evolved to solve problems efficiently, and that so happens to include problems of our own,' he told me. 'Animals and plants can efficiently capture sunlight to convert into energy, collect water where rainfall is negligible, make buildings with minimal materials, and so on. As the Earth's resources become increasingly limited due to overpopulation or climate change, biomimetics could become very useful to us.'

The structure of the glasswing's transparent panels has been used to inspire designs for more efficient solar panels. Montane phantom butterfly (*Pseudohaetyera hypaesia*), Colombia (TM)

Parker has investigated a wide range of species, including another kind of butterfly, the glasswing. I watched some of these insects fluttering through shady recesses in a Colombian cloud forest and was struck by how I could see through their wings, an evolutionary innovation that might help in the struggle against climate change. Parker explained why. 'The glasswing butterflies have transparent wings, but there's more to it than transparent materials. Window glass is transparent, yet its smooth surface reflects at least 10 percent of the sun's rays. The glasswings have evolved surface structures, each a hundredth of a hair's width, that serve to all but cancel out these reflections. This helps the butterfly to appear inconspicuous to predatory birds, but when placed on our solar panels they increase energy capture by 10 percent since more sunlight passes through.'

Parker enthused, too, about research into an insect that might help with the detection of fire. 'The *Melanophila* beetles seek out forest fires from up to 100 kilometers away in order to lay their eggs in recently burned wood. They do this by using tiny spheres beneath their wings, which resonate at the precise frequency of fire.'

Parker has also been working through his company, Lifescaped, to bring structural color applications to products, including paints for automobiles. 'This is the brightest color that exists and, since it involves microstructures rather than color pigments, it never fades. It will also replace some pigments that are currently mined by non-ethical or non-sustainable means,' he explained. One such source of inspiration in this innovation are hummingbirds, those iridescent avian gems.

And it is not only for color that we might look for this dramatic innovation, but also in relation to how such surfaces might require less intensive care. For example mud-shedding and water-repelling surfaces have been inspired by snakeskins, offering the prospect to create, among other things, building and vehicle exteriors that require less washing, thereby saving water and the need for cleaning chemicals.

The many different kinds of kingfisher that can be seen hunting along tropical rivers and in mangroves are equipped with the ability to enter water with minimum resistance. This biological design feature has been harnessed to render Japan's superfast bullet trains quieter, enabling them to travel faster, through copying the birds' bill shape. The reduction in air resistance that lowered the noise created by the trains also drastically cut the energy required to power them along at 300 kilometers per hour.

The possible applications arising from solutions invented by rainforest life are seemingly endless. Scientists are looking at ways to improve predictive software based on their evolution. Ant colonies are being studied to reveal ways of optimizing transport networks. Termite mound designs are being used to improve building energy efficiency. In engineering, designs inspired by nature are being developed in optics, acoustics, sensors and the development of composite materials.

How nature's genetic variety sustains us

Many medicines, too, were first identified among rainforest species. They include painkillers, treatments for infectious diseases, medications to prevent cardiovascular disease and anti-cancer drugs. Over 28,000 plant species have already been identified as having some medical benefit, and the more we look the more we find in the mega-diverse rainforests, where so many plants remain to be described and their beneficial properties to be fully understood.

Tom Prescott, a lead researcher at Kew Gardens, has spent more than two decades traveling through the rainforests of

Papua New Guinea. His research is focused on evaluating the effectiveness of local traditional medicines, and on developing low-cost screening techniques to help scientists in developing countries assess their own rainforest biodiversity for its pharmacological potential. He told me that 'several of the anti-cancer chemo-therapy drugs used by the NHS actually have their origins in plant chemistry. At Kew we have the expertise to isolate and identify the active molecules from plants.' He explained that 'Plants need chemicals to defend themselves from pathogens such as fungi that would infect them; after all, the plants can't just run away. Strange though it may seem we share a lot of genes with fungi, so our core biochemistry is surprisingly similar. So when a plant makes a chemical to kill fungi there's a good chance it will kill human cells too. This by itself isn't useful, but if by chance the chemical acts through a specific pathway or receptor that's unregulated in cancer cells, then that might be the starting point for a new drug.'

Prescott points out how the huge biodiversity in the rainforests translates into chemical diversity and that new technologies are allowing more rapid screening of plants and fungi, so long as they are still there to be studied. During his fieldwork he has lived with different tribes, asking what plants they use to treat different ailments. 'These tribes live in the middle of nowhere and their only access to medicine is plant medicine,' he says. He has been working with Papua New Guinean scientists to identify new research leads and is hoping to set up a clinical trial there to look at traditional cures for tropical ulcers.

Equally fundamental to human well-being is the contribution made by the tropical rainforests to our diet. Rainforests are the source of many of the world's most important staple crops, including rice, potatoes and maize—all of which are derived from wild rainforest relatives. Valuable timbers are another of the rainforest's evolutionary treasures to which huge financial values are attached. The diversity found in natural forest also provides benefits through the pollination of crops by wild creatures, and indeed the pest control services that insectivorous birds provide, in crops including tea, coffee and cocoa.

The genetic variety held in the millions of wild species found in the rainforests thus represents a massive asset, for engineering and design, our food supply and health and longevity. The potential practical benefits for humankind from rainforest species is by no means fully understood, not least because we haven't even named, never mind studied, most tropical rainforest species. But whether we've described it or not, that amazing natural diversity is only an asset for so long as it continues to exist. Given the rate at which some of the most diverse areas are being cleared and degraded, we cannot take for granted that everything of value will be there for us in the future. This is especially the case in relation to what are known as biodiversity hotspots.

Hotspots: where wildlife is richest and most threatened

As more data have been gathered, it has become apparent that our planet's biological assets are unevenly distributed, and the places where life is most unique are also very often places where deforestation and other manifestations of environmental damage have recently been intense. These areas, which are both biologically unique and under pressure, are called biodiversity hotspots.

To qualify for hotspot status a region must meet two strict conditions: first, it has have at least 1,500 unique higher plants, and second, it has to have lost at least 70 percent of the original area of natural habitat. On the basis of these two broad criteria, some thirty-four biodiversity hotspots have so far been identified. These together make up about 2.5 percent of the Earth's land surface and they are home to an astonishing 60 percent of the world's animal and plant species. Many of these hotspots contain particular centers of endemism, where the occurrence of unique wildlife is extreme. Some of these are areas that remained forested during the upheavals of the Pleistocene glaciations.

Professor Neil Burgess is Head of Science at the UN Environment's World Conservation Monitoring Centre (WCMC). He has spent decades working to conserve some of Africa's most biologically unique forests, especially in the moist uplands of Tanzania, and these experiences have given him powerful insights into the nature of biodiversity hotspots. At his office in Cambridge he told me about field visits to forests that had not previously been surveyed by ecologists, where 'in every single one we'd find new wildlife species—a chameleon, other lizards, a new species of frog, new snakes, plants, some small rodents, pretty much everything you want to think about.' Now and again Burgess and his colleagues would even find previously unknown primates: 'We came across a new monkey that ended up in its own genus.' The more data that was gathered, the more it became apparent just how diverse the mountain forests were, with each one very different from others.

'You find differences going up the mountain, which is sort of obvious with different altitudes and different temperatures, but you also find effects on the same mountain going sideways. You have river valleys going down a mountain and it can be the same altitude, the same forest, it can have the same tree composition and you go from one valley to the next and you can have two different species of frogs in the two different valleys which are inside the rainforest just separated by a ridge in the middle.' The implications of these kinds of findings for the conservation of wildlife species are important. 'Every time we look we go—OK, these areas that we knew were important are actually even more

Central America's species-rich highland rainforests are part of a
regional biodiversity hotspot. Talamanca Mountains, Costa Rica (TJ)

important than we thought. And when you take genetics into
account they are more important again. Whereas in lowland,
dried, burnt areas, of which there is a lot in Africa, you don't see
this kind of genetic variation across space, things are much more
the same over large areas. In these moist forests, though, over
just one kilometer they are genetically very different.'

The forests that Burgess and his colleagues have spent so
much time investigating are but one hotspot on that continent.
Others include the Guinea Forests of West Africa, those on
the mountains of Rift Valley in Uganda, Rwanda and the
Democratic Republic of Congo (DRC) and the highland forests

of Ethiopia. In Latin America the Atlantic rainforests, Central America and those in the Andes are hotspots, while in Asia the islands of Sundaland and the Philippines are among the super-diverse tropical rainforest areas.

These and other hotspot areas are not only places from which inspiration for new innovations in food, medicine, engineering and design might be found, as new species are discovered and those already known become better understood, but they can also be the source of genetic material from the wild relatives of species that are already economically important, especially in relation to our food supply.

The wild genes found among the relatives of cultivated varieties of rainforest plants—including maize, potatoes, cocoa, coffee, rice and avocados—might in future enable plant breeders to cope with new strains of crop disease, or help cope with the effects of climate change, through breeding drought-resistant varieties. In Britain, currently, advanced research is underway into breeding blight-resistant potato varieties, thereby sustaining yields while reducing the need for chemical protection, by moving genes from wild relatives into high-yielding domesticated varieties. This process is only possible when those wild relatives exist.

But even if we did somehow find out about all of the potential benefits hidden amongst the genetic mega-diversity of the tropical rainforests, the richness of life they hold in the future will depend on functioning ecosystems that sustain the millions of interactions going on between different life forms, the pollinators and herbivores, predators and decomposers, primary producers and parasites.

PART TWO
THE AMERICAS

U.S.A.

Gulf of Mexico

Bahamas

Turks & Caicos Islands

Yucatan Channel

Cuba

Bah'a de Campeche

Cayman Islands

Dominican Republic

Pue Rico

Mexico

Chiapas

Belize

Gulf of Honduras

Jamaica

Haiti

Golfo de Tehuantepec

Guatemala

Honduras

Caribbean Sea

El Salvador

Nicaragua

Los Angeles Cloud Forest

Panama Canal

Golfo de Urabá

Lago de Maracaibo

Costa Rica

Panama

Puerto Limón

Darién

Golfo Dulce

Drake Bay

Talamanca Mountains

Piedras Blancas National Park

Golfo de Panamá

Ve

Colombia

Río Caquetá

Chiribiquete National Park

Río Caquetá

Río

Galápagos Islands

Ecuador

Golfo de Guayaquil

Andes Mountains

Río Putumayo

Javari Valley

Andes Mountains

Acre State (Brazil)

Peru

Asháninka Reserve

Manu National Park

Otishi National Park

B

PACIFIC OCEAN

Atacama Desert

Rainforest extents: the lighter tint shows the 'original' pre-colonial extent of the rainforests, the darker tint what remains today.

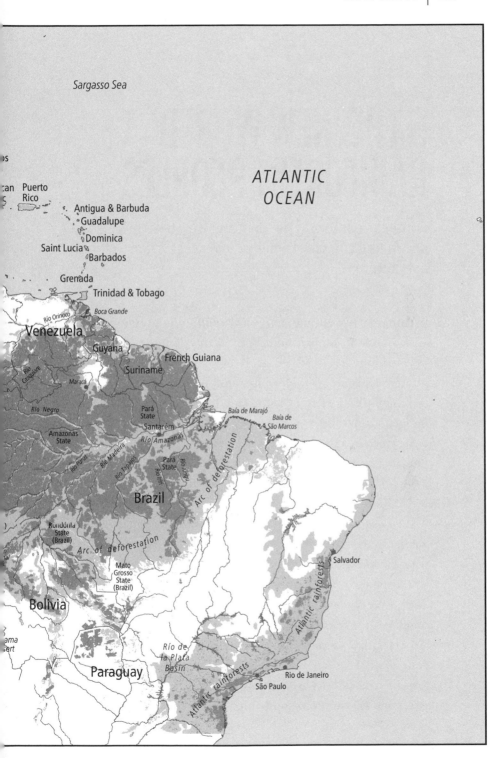

Sargasso Sea

ATLANTIC
OCEAN

Puerto
Rico

Antigua & Barbuda
Guadalupe
Dominica
Saint Lucia
Barbados
Grenada
Trinidad & Tobago
Boca Grande

Rio Orinoco
Venezuela
Guyana
Río Caquive
French Guiana
Suriname
Maracá
Río Negro
Pará State
Santarém
Amazonas State
Río Amazonas
Baía de Marajó
Baía de São Marcos
Río Purus
Río Madeira
Río Tapajos
Pará State
Río Xingu
Río Iriri
Arc of deforestation
Brazil
Rondônia State (Brazil)
Arc of deforestation
Salvador
Mato Grosso State (Brazil)
Atlantic rainforests
Bolivia
ama sert
Río de la Plata Basin
Atlantic rainforests
Rio de Janeiro
Paraguay
São Paulo

5 THE 'NEW WORLD' RAINFOREST PEOPLES

European contact in the rainforests of the Americas decimated complex societies and sustainable agriculture

A common perception of tropical rainforests is of wilderness areas free from human influences. If these wild jungles of our imaginations are populated, we're inclined to visualize bands of hunter-gatherers, with blowpipes and bows and arrows. While that might be a more or less accurate characterization of some indigenous rainforest peoples today, centuries ago many forest areas were in fact densely populated, and home to not only nomadic hunter-gatherer groups but also sophisticated societies living in and around cities. In the Americas the fate of both was sealed in the momentous year of 1492.

That was the year Christopher Columbus completed his first voyage to the 'New World' and his crew first made first contact with the forest-dwelling Carib Indians on the Caribbean island of Hispaniola (modern Haiti and Dominican Republic). They were evidently peaceful people, as Columbus himself observed in a message he wrote to his royal patrons in Spain: 'So tractable, so peaceable are these people that I swear to your Majesties there is not in the world a better nation. They love their neighbors as themselves, and their discourse is ever sweet and gentle and accompanied with a smile; and though it is true that they are naked, yet their manners are decorous and praiseworthy.'

Casual anthropological observation was one thing, Columbus's priority mission quite another. While noting the cultural and natural wonders he encountered in the New World, including the extensive forests full of novel wildlife, he also wrote about his real preoccupation: 'Should I meet with gold or spices in great quantity, I shall remain till I collect as much as possible, and for this purpose I am proceeding solely in quest of them.'

Columbus made three further voyages to the New World, on the last of which he reached the mainland of Central America, where he surveyed a substantial length of coast from present-day Honduras to Panama. En route and during the summer of 1502 his ships called at Isla Uvita, close by the the modern city of Puerto Limón. Columbus named this 'new' land Costa Rica (Rich Coast), from the gold jewelery worn by some of the indigenous people he came across.

Spain for a while had the lead in New World exploration and the race for gold and other riches, but other powers soon followed, with Portuguese, French and English adventurers intent on their share of fortune, fame and empire. On the Pacific coast of Costa Rica, on the other side to where Columbus landed, is a place known today as Drake Bay, named after the English adventurer, who dropped anchor there during his circumnavigation of the world between 1577 and 1580. I walked on the peninsula there in search of birds in the remaining tracts of primary tropical rainforest and it seemed to be one of the few areas of Central America that looks as it must have done to the first European adventurers. Such impressions are wrong, though—as they are for the forests that remain along the coast visited by Columbus, or indeed nearly all the other New World tropical rainforests that remain intact, for while these rainforests still retain most of the frogs, flowers, lizards and birds that would have been seen by early explorers, most of the people are gone.

New World agriculture

When European explorers first sailed along the coasts of Central America and probed the great rivers of South America, human societies had been established there for millennia. A few thousand people are believed to have crossed into the Americas from Siberia about 16,000 years ago and gradually colonized all the way south to Tierra del Fuego, including through the rainforests of Central and South America. While at the end of the fifteenth century the indigenous people of the New World rainforests didn't possess many of the technologies available to Europeans, many communities did have advanced agricultural practices. Evidence of farming in the Amazon basin (believed to be the first in South America) dates back about 8,500 years ago, with the cultivation of, among other things, manioc, beans and chillies. Indeed it was the indigenous peoples of the Americas who were the first to domesticate many crops that are vital to the functioning of our modern world, providing far more long-term practical value than the gold that was so actively sought by the conquistadors.

One hugely valuable rainforest tree grown by indigenous people and soon seized upon by Spanish explorers was cocoa. The Olmec Indians, who lived among the tropical rainforests of what is now southern Mexico and Guatemala, were probably the first to consume a chocolate-like food, about 3,000 years ago, most likely in a drink. They cultivated cocoa trees, as did the Mayans and the Aztecs, and by the time Spanish travelers arrived chocolate was embedded in marriage rituals and religious rites. Today cocoa is grown right around the tropical rainforest belt in warm and wet lowland areas within about 10 degrees of latitude of the equator.

More than 10,000 years ago American rainforest Indians were also the first people to eat avocados, initially collecting wild fruits before cultivation started about 5,000 years ago. Pineapples are a kind of bromeliad and come from the world's bromeliad 'capital': Brazil. They appear to have first been domesticated by Indians living the Atlantic rainforests in the south of that country and in adjacent Paraguay. By the time Europeans arrived in the New World pineapple cultivation had spread to Central America and the Caribbean and since then has been grown right across the wet tropical regions. Even more integral to world agriculture is the tomato—unknown to Europe until the sixteenth century—whose domestication first began in the Andean region around 1,300 years ago.

Then there are the global staples that have underpinned human population growth. The tall maize plant we know today is the result of domestication in Central America about 8,500 years ago, when people began to selectively breed a kind of teosinte grass, probably in a lowland forest environment. Modern maize, with its big head of tightly packed yellow kernels, today produces 900 million tonnes of grain. Potatoes, too, were a New World crop, domesticated by rainforest people 8,000 to 10,000 years ago. There are more than 150 species of wild potatoes, occurring across the Americas, growing in conditions that range from dry desert along the Peruvian coast, to the inter-Andean valleys, at altitudes of 4,200 meters above sea level. The highest concentration of species is in Peru and Bolivia and it is to there that the ancestors of most modern species can be traced, brought to Spain by conquistador travelers and to England by Sir Francis Drake, who had them on board the *Golden Hind* as he headed north along the coast of Central America.

The indigenous rainforest societies of the Americas not only identified these crops and introduced farming systems to exploit them, but they did so on a scale and with a complexity capable of supporting substantial cities. We have long known of pre-Columbian cities from the ruins of Mayan, Aztec and Inca civilizations. What is less known—but which may have important lessons for modern times, and in particular for carbon storage—is how their civilizations were able to maintain fertility in rainforest soils.

Soils that sustained cities

The Spanish conquistador Francisco de Orellana led the first party of Europeans along the Amazon in 1542, going downstream from Ecuador. According to Gaspar de Carvajal, his official chronicler, there was 'one town that stretched for fifteen miles without any space from house to house, which was a marvellous thing to behold,' and where the Rio Tapajos joined the vast main channel of the Amazon their expedition found the banks swarming with people. Inland 'could be seen some very large cities.'

Recent archaeological research has backed up these accounts, revealing large-scale organized societies from pre-Columbian times. As deforestation has spread, earthworks dating back 1,000 to 2,000 years have been revealed, showing how large areas were once under settlement; more than 300 have been discovered in the Brazilian states of Acre, Amazonas and Rondônia, with about 70 more in Bolivia. Satellite imagery and earth-density scanning have led to the discovery of more and more 'lost cities of the Amazon'—dense, riverside urban centers occupied by up to 10,000 people. Agricultural areas and orchards stretched for tens of kilometers along the riversides and include elevated fields that kept crops clear of the seasonal flooding.

Considering what we know about the vulnerability of tropical rainforest soils, the fact that these people could grow enough surplus food to support such a level of urbanization was a

Extensive areas of *terra preta* – rich dark soil – found in the Amazon basin are of pre-Columbian human origin (Manuel Arroyo-Kalin)

remarkable discovery. But it turns out that the rainforest soils had been replaced with man-made ones in large swathes of pre-Columbian farms. The people had developed sophisticated gardening methods that dramatically increased and then maintained fertility. The improved soils produced a huge quantity of food—far more than hunting and gathering in the forest could yield.

These fertile soils can still be found along the banks of the Amazon today and are estimated to cover as much as 150,000 square kilometers. They are darker than the typical ferrous rainforest soils and were deliberately created to increase food production. They are known as *terra preta*, or dark earth, and are the result of centuries of composting and nutrient enrichment. Pre-Columbian farmers added a mixture of unused food remains, bones and human excrement, charcoal and wood ash to the soil. As a result, the level of dark organic material—the humus—was increased, feeding the microbes that release the nutrients that enable plant growth. Even though such soils are upwards of 2,000 years old they remain today rich in nitrogen and phosphorus, and have a very high carbon content.

Numerous fragments of clay pots found in the *terra preta* suggest that organic material was systematically collected and composted for application to the farmed plots. It seems each person had their own clay vessel in which they deposited their excrement. A layer of charcoal powder was added each time the pot was used, with a lid to control smell and the risk of disease.

The charcoal, created through subjecting organic materials such as crop wastes and wood to high temperatures in the absence of oxygen, was a key component and manufactured on a large scale. Pre-Columbian farmers applied it to their gardens, mixed with composted organic matter, because it helped increase yields. This was because, in addition to increasing microbiological activity, it enhanced soil moisture retention. In our modern world adding organic matter to soils in the form of biochar has become of great interest, because of its potential to store huge quantities of carbon in agricultural soils.

Into these dark earths tall fruit- and nut-bearing trees were either planted or left when the forest was cleared for farming. A shrub layer below produced bananas and avocados, while on the ground were a variety of crops, including the 'three sisters' typically grown together in traditional tropical New World farming—beans, maize and squashes—as well as potatoes and chillies. The tall trees protected the soil from heavy tropical rains that would cause soil loss and nutrient leaching, while the layers of vegetation mimicked the structure of the forest, keeping the humidity high and increasing the overall yield of food. The diverse mix of crops meant that pest invasions were less likely and the complexity encouraged insectivorous birds, further helping to keep insect pests under control.

Much tropical farming today wrecks soil by causing erosion, compaction and depleting soil organic matter, but these pre-Columbian farmers were improving it for agriculture, enabling the rise of cities along the banks of the Amazon. These dark soils have been described as a lost legacy of the ancient inhabitants of the Amazon but modern farmers continue to farm ancient black earth soils and even now enjoy higher yields.

The pre-Columbians also managed the forest vegetation that surrounded their garden cities. Recent studies suggest that some 10–12 percent of the Amazon's terra firma forests (the ones beyond the extensive flood-plain *várzea* forests of the main river channels) are in part the result of ancient management. Plantations of useful native trees were established, and some species effectively domesticated: fruit trees were sited to attract animals such as pigs, which were hunted for meat.

I saw some of this legacy of indigenous people's influence on the forests in Central America, where centuries ago people had planted fruit trees on the tops of hills where pigs could be more easily surrounded, ambushed and slaughtered. Many old mountain almond trees found on forest hilltops today are thought to have originally been deliberately planted as magnets for pigs and agoutis. Other trees were planted to produce fruits for human consumption.

The different species of palms that dominate the ground layer in many New World rainforests were vital to the economies of the indigenous people. The leaves of some are good for cooking in—the equivalent of foil. Others are bendable and can make handy drinking cups. Some have leaves that are good for roofing, while the trunks of others make excellent curved

New World rainforests are rich in palm species, many of which are utilized by forest peoples. Los Angeles Cloud Forest, Costa Rica (TJ)

planks for boat-building. Other palm species have stems that are used to make the blowpipes through which darts tipped with frog poison were used in hunting. Plants and animals were also widely drawn upon for medical purposes.

We have no census data on the population of the indigenous populations of the rainforests when the first Europeans arrived in the Americas, but it is estimated that the Amazon alone had a population of some 8 million people. These societies had developed over many thousands of years and were integral to the natural systems that sustained them. It is increasingly apparent that they weren't (and those that remain aren't) passive recipients of rainforest productivity, but active shapers of the lush green world they inhabited.

Good morals

Having established that a large number of people were living in the new lands, the Spanish conquistadors were ordered by the Pope to spread Catholicism among the indigenous peoples. In 1493 an edict signed by Alexander VI ordered the Spanish conquerors thus: 'We command you to instruct the aforesaid inhabitants and residents and dwellers therein in the Catholic faith, and train them in good morals.'

While there was a paternalistic religious dimension to some of the interactions that would follow, Columbus set the tone as to how relations with the indigenous peoples were to unfold. 'They do not bear arms,' he wrote, 'and do not know them, for I showed them a sword, they took it by the edge and cut themselves out of ignorance. They have no iron. Their spears are made of cane . . .' Having noted their vulnerability, he added that, 'They would make fine servants' and 'with fifty men we could subjugate them all and make them do whatever we want.'

Michele de Cuneo, traveling with Columbus on his second voyage, recorded the following experience. He'd been given a 'very beautiful Carib woman' captured by Columbus and wrote that 'When I had taken her to my cabin she was naked—as was their custom. I was filled with a desire to take my pleasure with her and attempted to satisfy my desire. She was unwilling, and so treated me with her nails that I wished I had never begun. But—to cut a long story short—I then took a piece of rope and whipped her soundly . . . and we came to terms.' So much for papal edicts demanding that religious teaching and morality should shape the new relationship between the Europeans and the forest peoples they'd found.

The chief motivation, however, was to subjugate and dominate for profit and power. The peaceable forest people found themselves confronted by alien societies who had guns, swords and ocean-going ships. Their leaders were driven by fortune, fame and royal favor and they were highly motivated to serve their own ends with little regard for the consequences. What followed was, for the indigenous peoples of the New World rainforests, nothing short of a catastrophe.

Murder and enslavement caused massive loss of life and suffering, but far worse were the diseases the Europeans brought with them, to which the rainforest peoples had no resistance. The impact of these diseases, often not especially harmful in Europe, was in part down to the Indians not having domesticated animals like cattle and horses. Europeans had been exposed to the pathogens carried by such animals for thousands of years and had derived resistance to certain infections. John Hemming, who has written perhaps the definitive history of Brazilian Indians, told me that the overriding reason for the population collapse among the rainforest peoples was disease: 'Smallpox, influenza, measles and especially pulmonary diseases were most important and every single time the pattern was the same. The scale of what happened was devastating. The Jesuits, who were in Brazil by the mid-1500s, sent letters back to Rome to provide updates on baptism and so on. From their base in Salvador they reported how in five years all the indigenous population had gone. In the 1560s they traveled along the coast of Brazil and found village after village totally dead. It wasn't just along the coast but far into the forests, too.'

From a distance of more than five centuries it is impossible to precisely quantify the loss of lives to disease, but the fragmented sources that do exist suggest that in some populations more than 90 percent of indigenous people died. Cultures that had evolved in the rainforests over millennia were wiped out within a few decades. And with them went the knowledge of how to farm the forest without destroying it. In the place of traditional wisdom, a new mindset had arrived. Instead of regarding the intact rainforests as a precious source of livelihood, the incomers saw it as a hostile wilderness from which to extract timber, slaves, rubber, gold, spices, meat and animal skins.

It was not only in the Amazon that the destruction of the native cultures took place. The Andean region, one of the first places to be colonized by the Spanish (from the 1530s), lost about three quarters of its indigenous population through European diseases, war and slavery. In the Andean region the diseases the Europeans brought to the Americas got there before even the Spanish themselves, traveling through Central and South America from the Caribbean. The conquistadors not only killed off the people but destroyed entire civilizations, including the Incas, whose cities were sacked in pursuit of gold and other precious metals that were to underpin the rise of Europe's new capitalist economies. The indigenous people who lived in Brazil's southeastern coastal rainforests were also quickly decimated.

When the Portuguese first arrived in 1500 the dominant group there, the Tupi, might have numbered as many as one million. Not for long.

Through the 1600s and 1700s Indians were driven into slavery in their droves and with an appalling loss of life, even before they arrived at the plantations and ranches where they would be forced to work. Some were sent into the forests to gather commercially valuable products while their families were kept hostage to make sure they came back. Others were forced into construction work, including the building of ocean-going ships, while more were put to work in brick and tile factories. Slavery was in theory banned in Brazil in 1748, but the practice continued, and on a grand scale, especially in remoter areas such as the Rio Negro region in the northern Amazon. As slaves died from overwork, disease and malnutrition, so more slaves were rounded up from forests along ever more far-flung tributaries, taking disease deeper and deeper into the rainforest.

In 1491 the human population of the rainforest lands of the New World was 100 percent indigenous people and several millions strong. Today the Brazilian Amazon is estimated to be home to between 280,000 and 350,000 indigenous people, with about 180,000 living 'traditionally.' Overall, the indigenous proportion of Brazil's current population is just 0.4 percent. It is no exaggeration to talk of genocide—and one that continued for centuries.

The rubber boom

Among the discoveries made by early explorers was rubber, which had been used by indigenous people for thousands of years before European contact. Rubber balls were part of a ritual game and Columbus brought some back from the West Indies on his second voyage. The latex from which the balls were made was noted as possessing excellent waterproofing properties, as well as erasing pencil marks from paper. But it was not until the nineteenth century, following the refinement of industrial techniques, that rubber became an industry in Europe and the USA, with the large-scale manufacture of waterproof rubber shoes and clothing. Then in the 1880s, with the invention of the automobile, it became a boom industry, with ever-increasing quantities required for tire manufacturing.

In 1876 Sir Henry Wickham famously transported rubber seeds from the Amazon to Kew Gardens—from whence rubber plantations were created in the British colonies of Ceylon and Singapore. However, most nineteenth-century rubber came from wild trees growing in the Amazon rainforests. There, European and American rubber barons enslaved hundreds of thousands of people from remaining Indian tribes to work in new plantations. Slave masters of African descent were imported from the Caribbean to help manage the labor force. These *muchachos* kept order with fear born of extreme violence and were periodically instructed by the white rubber plantation owners to make raids on Indian settlements to round up new recruits.

The captured Indians were sent far into the forest to find rubber trees. They made deep incisions into the bark in order to get every drop of latex, often cutting so deep as to kill the tree. They waited a few days for the rubber to harden, cleaned out impurities such as twigs and rolled it into balls for transport. The Indians were set quotas for rubber harvesting for which they were theoretically paid, but in reality they were trapped in a system of debt bondage with impossibly demanding targets. Many died from overwork, others from the effects of flogging, or being locked in stocks without food or water.

It is estimated that the brutality and enslavement visited upon the indigenous Amazon population by rubber barons reduced

Enslaved Amazon Indians during the rubber boom of the early 1900s, from Walter Hardenburg's 1912 book *The Putumayo, the Devil's Paradise; Travels in the Peruvian Amazon Region and an Account of the Atrocities Committed upon the Indians Therein*

their population by as much as 90 percent. Among those that survived were groups who fled into the forests. Many of the isolated Indian groups that remain today in the western Amazon, in Peru and Brazil, are descended from such survivors. For more than a century they have lived deep in the forest, many of them entirely shunning contact with the outside world.

Survival International, the UK-based group that campaigns for the interests of indigenous people, suggests that many tribal groups who today remain 'uncontacted'* are descendants of survivors of atrocities committed long ago during that rubber boom, and that this has directed their life in isolation. One can only imagine the stories that fuel their cultural history, of what happened to their peoples long ago when the white men and their black slave masters arrived, and why interaction with the outside world must be avoided at all costs. A famous photograph taken from a helicopter in the Envira region on the Peruvian–Brazilian border in 2008 showed rainforest Indians with skin painted bright red and with partially shaved heads standing in a clearing next to their houses. Arrows drawn back in bows aiming into the sky and toward the machine roaring above them, their message was very clear: 'Keep away.'

And it was not as if the world had issued an apology and changed its ways. A notorious incident indicative of the ongoing catastrophe took place in the upper reaches of the Aripuanã river during 1963. It became known as the 'massacre of the 11th parallel.' The head of a rubber company decided that the Cinta Larga Indians were hampering the profitability of his company.

---

* The term 'uncontacted' is used of isolated tribes in Brazil, whereas in Peru the tribes are referred to as living in 'voluntary isolation.' Survival International doesn't use 'voluntary' because it implies that uncontacted tribes have the luxury of choice, whereas the reality is that they have opted for isolation as a strategy for survival.

'These Indians are parasites, they are shameful. It's time to finish them off, it's time to eliminate these pests,' he was reported to declare, before hiring a plane and dropping dynamite on an Indian village. He then made arrangements for gunmen to finish off the survivors. It was far from the only such crime, and local murders of indigenous people, often committed by groups of miners, are still reported.

And, as Catherine Caufield reported, in her groundbreaking 1984 book *In the Rainforest*, the missionaries were little better. First unleashed to spread Christian teachings and values among indigenous groups by the Catholic Church in the fifteenth century, they persisted with their mission for hundreds of years. Caufield interviewed US missionaries from the Florida-based New Tribes Mission who were 'hunting down' the last nomadic group in Paraguay, the Ayoreo. Once captured, the tribe would be put in camps for Christian conversion and there, deprived of 'jungle food,' and exposed to infection, most would die, just as they had following first contact with the conquistadors. The missionaries were part of an unholy modern alliance that included US companies prospecting for oil and uranium and the Paraguayan government.

Survivors

Since 1900 more than ninety Amazon tribes have been wiped out. However, despite the continuing depredations, there are still groups of indigenous peoples living in the remaining areas of New World tropical rainforest. In Mexico and Central America tribes include the Achi,' Tepehuán, Maya, Ixcatec, Miskito, Pech, Q'eqchi,' Rama Nicaragua, Seri, Taino, Yucatec and Zapotec. In South America the best known are the Yanomami of the Amazon rainforests, while others include the Asháninka, Bororo, Chayahuita, Emberá, Enxet, Kuna, Makuxi, Nukak, Secoya, Tupi, U'wa, Yora, Wichí and Wayuu. There are, too, a significant

An uncontacted tribe, sometimes referred to as the Uncontacted of the
Rio Xinane, in Acre state, Brazil. Bows are drawn as a helicopter hovers
overhead (Gleilson Miranda/FUNAI/Survival)

number of groups who live in isolation from the outside world.
John Hemming believes that in Brazil alone there are many as
twenty 'uncontacted' groups.

As we'll see later, the fate of the forests is very much bound
up with the destiny of these remaining indigenous societies,
rendering ongoing efforts to save the rainforests an endeavor
that is as much a social project as it is an ecological one. This
social dimension is not only about tribal peoples, though. Other

societies that are not descended from indigenous people depend on the forests, as well. Groups first derived from slaves, for example, live in and adjacent to forests, and are often dependent upon them, in Colombia and parts of Central America and Brazil. Then there are the poor people who live around the edge of the remaining forests, some of European rather than indigenous origin, and who in some places are the reason why it continues to be cleared to make way for more farmland. Many of these people are refugees from conflict, or landless people seeking a living, or people deliberately moved to forested regions by government-backed resettlement programs.

But, for all of these people, the pressures have altered during recent decades. Indigenous peoples are now less affected by direct persecution, but increasingly touched by the physical destruction and degradation of their forest homes, including by logging and clearance for farmland.

6 FOREST CLEARANCE IN THE AMERICAS

The exploitation of 'virgin terrain' and the unfortunate mirage of rainforest fertility

By the late nineteenth century the idea of the Amazon, and other rainforests, as empty, wild and free of human influence had more of an element of truth. As we have seen, whole civilizations had been wiped out, and indigenous people reduced to a fraction of their populations. Theodore Roosevelt, twenty-sixth President of the United States, and an enthusiastic big game hunter in South America, declared of Amazonia: 'Surely such a rich and fertile land cannot be permitted to remain idle, to lie as a tenantless wilderness, while there are such teeming swarms of human beings in the overcrowded, overpeopled countries of the Old World.'

It was a sentiment shared by most governments of Latin America, and indeed beyond, throughout the twentieth century. And it is still directing policy in much of the region, as governments sacrifice the forest environment and its peoples to the short-term economic opportunities of energy, mining, logging and farming. In the frontline of the clearance were Brazil's coastal Atlantic rainforests. I visited there shortly before taking on my role leading the rainforest campaign at Friends of the Earth and saw for myself the scale of devastation that had resulted from hundreds of years of degradation.

American clearances

Beneath low grey skies, cools mists and drizzle, I headed with Brazilian colleagues north out of Rio de Janeiro and into a landscape that although once the domain of rainforest Indians had over the course of half a millennium been effectively 'Europeanized.' Mixed farms produced fruits and vegetables, while larger ones cultivated soya beans and sugar cane on vast fields that extended to sprawling flat horizons. There were cattle pastures, plantations of trees destined for pulp mills and coffee estates. Odd patches of native rainforests remained, particularly on steeper slopes, but they were few and scattered.

In 1500, when the first Portuguese ships navigated the nearby coast there was an unbroken band of rainforests extending along more than 3,000 kilometers from south of modern-day São Paulo in the south and north to around the modern port of Natal. At its most northerly extent it comprised a narrow coastal band a few tens of kilometers deep, while the southern coastal forests extended much further inland to Paraguay and Argentina.

Separated from the great block of rainforest in the Amazon basin by the savanna woodlands of Brazil's vast interior and drought-prone northeast, the parlous state of South America's Atlantic rainforests was the result of a long history of exploitation. Before being subjected to large-scale clearance by fire and axes these forests covered a total of about 1.2 million square kilometers—nearly a quarter of the extent of the Amazon rainforests. But they'd been under mounting pressure since the Portuguese colonial period. First trees had been cut and then the forests supplied charcoal for cooking and smelting metals. The fields were opened for cattle and crops, while in the hills much of the forest was replaced by ranks of coffee bushes.

By the time of my visit, in 1990, more than 85 percent of the original Atlantic forest had gone, giving way not only to farmland but latterly to urban areas, including two of Latin America's largest cities—Rio de Janeiro and São Paulo. Most of the 15 percent that remained was regrowth, with only about 1 percent of original forest still more or less intact.

Logging, charcoal production and the establishment of plantations by the Spanish colonists took a similar early toll on the rainforest that clothed the slopes of the Andes, giving way to farms and pastures. Traveling in the Atlantic forests of Brazil during the 1990s and the highlands of Colombia, Ecuador and Peru a decade later, I was struck by how few fragments of forest in these regions of early deforestation had survived.

Across the Darién Gap and into Central America a similar story of deforestation unfolded. In pre-Columbian times a contiguous band of rainforests ran from southern Mexico, Guatemala and Belize south to Panama, but this is now mostly cleared or highly fragmented. In Mexico tropical rainforest once clothed much of the southeast of the country, but today little survives outside of the region of Chiapas, on the border with Guatemala. The destruction of all this forest began after the arrival of the conquistadors, with logging and clearances for cocoa plantations. But, as Theodore Roosevelt observed in relation to the Amazon, even the forest that survived into the twentieth century was regarded as empty wilderness fit for conversion.

In the post-war period deforestation became more rapid, not least because of the invention of the diesel-powered bulldozer and chainsaw, adding to what could be done with axes and fire. Between 1950 and 2002, Guatemala lost half of its forest cover, mainly due to conversion to cattle ranching and oil palms. Drug money played a part, too, in Guatemala—as it does in many South American rainforest areas—with logging businesses set up to launder the proceeds of trafficking. Another strong driver of

Central American forest loss has been the concentration of land ownership in relatively few hands, pushing landless people ever further up the hills and into the remaining forests.

In 2001 I visited nearby El Salvador, whose countryside has very little remaining natural forest. The history was similar to that of Guatemala. Forest clearance began during the Spanish colonial period, especially in the 1840s, when there was a major expansion of coffee plantations, while a concentration of land ownership caused pressure on small farmers to deforest new areas, occupying less fertile lands, much of it located on hillsides. El Salvador was once largely covered with dense rainforests but today less than 2 percent of the country has primary forest.

The forests in neighboring Honduras met a similar fate, where the expansion of large agricultural estates during colonial times was followed by a rising population of small farmers seeking more land upon which to grow food to feed themselves and local markets. The process continued into modern times, and from 1990 to 2005 that country experienced the highest deforestation rate in Latin America, losing about 37 percent of what was then left. By 2015, primary forests covered just 13 percent of the country. Honduras has also been a country where those campaigning for the protection of the remaining forests risk their lives by doing so. During recent decades hundreds of people have been murdered because of their conservation work, more than in any other country, with 116 environmentalists killed in 2014 alone.

To the south, Nicaragua lost most of its forests to cattle ranches, logging and forestry plantations. As elsewhere in Central America, encroachment into forests by small-scale farming has been a major driver, also in large part ultimately down to landlessness. This is a large country, however, and with about 15 percent of its territory still supporting extensive

A fragment of subtropical highland forest in the Colombian Andes lies next to a cattle pasture and a potato field (TJ)

primary rainforests it has some of the biggest blocks that remain in Central America.

In Nicaragua the disparities of land ownership were not only a significant cause of deforestation but also a flame in the powder keg of social tensions that exploded into civil war. The same was the case in the late twentieth-century civil conflicts that caused such terrible bloodshed and misery in El Salvador (during the course of which most of the remaining forests were decimated) and also Guatemala, where a thirty-six-year-long civil war left over 200,000 dead (83 percent of them indigenous people).

One more country where there was a similar historic pattern of forest loss, but during recent decades far more positive outcomes, is Costa Rica. The expansion of coffee led to widespread rainforest clearance during colonial times, as did the production of sugar cane. Then, during the 1970s and early 1980s, rampant deforestation got underway, mainly because of the expansion of cattle ranching. Driven in part by the so-called 'hamburger connection,' beef production was increased to feed the United States' fast-food boom. Then government policy changed to become far more pro-forest and because of that the rate of deforestation not only went down but since the 1980s Costa Rica's forest cover has increased, as efforts to protect and restore native forests have delivered results. Many species of wildlife have begun to recover too, in part due to a ban on hunting and collecting. Significantly, the economy has grown at the same time. We'll return to Costa Rica again later to find out why and how that became the case.

To the east of Costa Rica lies Panama—a focal point for the Spanish colonial quest for gold and natural resource extraction. Panama, however, held onto a substantial proportion of its original rainforests until the second half of the twentieth century, when half of its then remaining primary forests were wiped out, largely to make way for farms and cattle ranches. Nonetheless, about 40 percent of the country remains covered by old natural forests, including a substantial expanse in the west of the country that forms part of a cross-border National Park with Costa Rica.

The rainforests on the islands of the Caribbean were peculiarly vulnerable to colonial-era clearance, as confined and accessible territories, and were easily logged. Soon after their discovery by Europeans there was very little natural forest left, and today there is almost no primary forest. Hispaniola, the second largest island and the one first reached by Columbus, was soon mostly deforested. Colonists took the most valuable hardwood trees to make ships and buildings, while less sought-after species were used for fuel. What was left was burnt to make way for plantations, mainly for sugar cane. Cuba, the largest island, was in 1492 more than 90 percent forested but by 1900 this had shrunk to about 5 percent. A similar story unfolded across the other major islands, including Puerto Rico and Jamaica. A few fragments of forests escaped the process of land clearance, and many are today prized as National Parks, but they are few and far between.

So it was that even before the French Revolution many New World forests were eliminated. The main reason was logging, followed by plantation agriculture to produce crops, including sugar—the first tropical commodity grown in large monocultures. It began with the Portuguese leveling Brazil's coastal forests and the Spanish colonists deforesting the Andean highlands, the Caribbean islands and much of Central America. From the 1830s onward coffee became a key Latin American export and took up much land in Brazil, Colombia and Central America. The pressures continued across the centuries and persist today, where there remains any forest left to clear.

The Orinoco and the Amazon

The biggest continuous tracts of New World tropical rainforest lie across the vast river basins of the Orinoco and Amazon.

The Orinoco flows in a giant arc from the Colombian Andes through Venezuela to the Atlantic Ocean, draining an area larger than Texas. With its flooded forests and great diversity of plant and animal life, it is one of the most intact major rainforest river basins in the world, although it faces mounting pressures. Large areas of the flooded forest have already been cleared for crops and cattle pastures, while mining projects and hydroelectric dams threaten the river and its forests.

Far more extensive still are the rainforests of the Amazon. Historically the pressures on the Amazon (as opposed to its people) were relatively light, though continuing global demand for rubber in the first half of the twentieth century began to impact forest cover. It included the establishment between 1928 and 1945 of a one-million-hectare plantation by the Tapajos river in Brazil—the brainchild of Henry Ford to secure resources for his Detroit automobile production lines. Ford's plantation was the first large-scale monoculture established in the Amazon, and unlike earlier rubber ventures was run with relatively enlightened worker welfare. However, it was to prove a doomed venture. The rubber trees that grew naturally in the forests (and which had hitherto been the main source of rubber from that part of the world) didn't do well in the huge plantations, where they fell victim to disease infestations.

Despite these early attempts at industrial farming it was not until the 1970s and 1980s that extensive deforestation took off in the Amazon. The process was hastened through a deliberate resettlement strategy that involved building major roads into areas of once remote forests, with financial assistance provided by international development agencies, including the World

Bank. Poor people from the impoverished and drought-prone northeast of Brazil were moved into the western Amazon, where they cleared land for crops, bringing pulses of forest loss in their wake. Satellite photographs in the years that followed the upgrading of the infamous BR-364 highway revealed the development of a 'herringbone' pattern of deforestation, as parallel 'ribs' extended out from the 'backbone' of the main road and deep into the once undisturbed primary forest.

In remoter parts of the western Amazon, including the lowlands of Colombia, Peru and Venezuela, the pressures during the 1980s and 1990s were lighter. The economies of these countries were more centered on the Andes, and with limited road and river access for extracting natural resources the pressure on the lowland forests was less severe. To the south in Bolivia, however, there was significant deforestation in some parts of the Amazon lowlands, and in drier forest areas too.

Recent analysis reveals how in these remoter parts of the Amazon, where deforestation was until recently quite light, things have changed fast. Illegal logging and the establishment of commodity crops, such as palm oil and cocoa, are among the reasons. So is the cultivation of coca, the plant from which cocaine is made. Grown on the moist lower Andean slopes of the Amazon in Colombia, Peru and Bolivia, narco-cultivation has posed an increasing threat to the integrity of the remaining rainforests there.

In addition to deforestation caused by small-scale farmers, larger land interests have also played an increasing role over recent decades, opening extensive areas for cattle ranching, burning the forest during drier periods to replace trees with pasture and establishing massive soya plantations. There are, too, more and more incursions into remote areas of forest, including protected areas, by loggers searching for gold, and for valuable mahogany trees.

The combined effect has been to create, from the 1980s onward, an 'arc of deforestation' across the southern and western parts of the Amazon basin rainforests, extending for thousands of kilometers from the Atlantic to the Andes. This continent-scale deforestation is now clearly visible from space.

Big beef in the Amazon: the soya clearances

As elsewhere, the modern destruction of the Amazon's rainforests begins with logging for high-value timber species, followed by clearance for farming. It has been happening on an ever-larger scale driven by the demands of increasingly rapacious international commodity markets.

In the wake of timber extraction, cattle pastures were established to produce beef and leather, much of which went right around the world to end up in meat pies in British and American supermarkets, or on the upholstery of high-end cars. The ranchers detected rising demand, and in response to powerful market signals cleared more land. Across the southern Amazon cattle grazing on thin pastures between the charred stumps of rainforest trees became an increasingly common sight.

Expanding demand for livestock not only showed up in bigger markets for beef and leather, but also for animal feed. Fast-growing markets for protein-rich animal feed began to be supplied with soya, grown on a truly vast scale. Soya was originally grown in Asia before becoming adopted as a major crop in the United States, and then spread to the savannas of central Brazil—the Cerrado woodlands. Large-scale clearance of this biologically diverse ecosystem, with its anteaters, rare parrots and unique plants, caused widespread ecological damage.

Then the development of new soya varieties permitted the cultivation of this crop in the more humid conditions of the Amazon. This included the rapid and now massive expansion of

Land deforested for soya cultivation, Mato Grosso, Brazil (Greenpeace)

varieties genetically modified to withstand herbicides designed to kill all other plants. South America (not only Brazil but also Argentina and Paraguay) is now the world's largest source of soya, and as market demand for pork, chicken and beef continues to rise, so does demand for the soya-based animal feed needed to sustain the industrial meat and dairy factories in the United States, Europe and China.

To facilitate the shipment of increased soya output from the Amazon to global markets, a new port was opened in 2006 near

Santarém in the Brazilian state of Pará. This part of the eastern Amazon had for a long time been a hotspot for deforestation and the upgrading of infrastructure there was set to make things worse. In addition to the expanding port facilities, the bumpy dirt roads were hard-surfaced to enable easier movement for the large trucks and machinery needed to expand production. The port became a magnet for farmers wishing to cash in on the soya bonanza who were incentivized by the soya traders with capital costs to get production up and running. Many came from the south of Brazil, where land was more expensive. By selling a farm there, they could afford to get a much bigger one in the Amazon, and through being connected to global markets via the new infrastructure could make a lot of money.

During my first visits in the early 1990s Brazil was still very much a developing nation. Going there again in 2011 and 2014 I found a country transformed. Many more people lived like Europeans and Americans. Roads were packed with new cars and poverty was less evident, at least in the smarter downtown districts of São Paulo and the capital, Brasilia. Although great disparities remained between rich and poor, the economic strategy seemed to have paid off, at least in narrow GDP growth terms. And the overall increase in national wealth was in large part down to the increased output of agricultural produce, especially of beef, soya and sugar.

While much of the emphasis in efforts to conserve the tropical rainforests between the 1970s and 1990s was focused on the impacts of slash-and-burn farming and the logging industry, the main threat was increasingly large-scale agriculture, often feeding global markets (rather than people living in and around the places where the forests one stood). One 2012 study estimated that commercial agriculture accounted for about 40 percent of recent deforestation across the world. The percentage varied by continent but was highest in Latin America, where some two thirds of forest loss was down to such land use change (in Asia and Africa it was more like one third). As time has gone on, the proportion of forest loss arising from large-scale commercial farming has increased and, when the impacts of smallholders are included, the proportion of all deforestation that is caused by agriculture is more like 80 percent. The UN Food and Agriculture Organization estimated that, between 2005 and 2010, about 2.8 million hectares of forest was lost in Brazil (an area larger than the US state of Massachusetts).

Having said this, there is some small cause for optimism. While the overall global rate of tropical deforestation continued to increase year on year, Brazil eventually bucked the trend. The country was still losing forest but at a slower rate—forest loss in 2012 was about 70 percent less than that of 2004. It was a positive step, but, at the time of writing, under President Temer, the trend is not currently going in the right direction.

7 COUNTING THE COST: GREEN SHIELDS

The effects of deforestation range from droughts to increased vulnerability to extreme storms—we are in an era of consequences

In September 2014 I traveled to São Paulo with Cambridge University colleagues to work with a group of business executives. Our brief was to help them explore challenges posed by environmental change, and to design business strategies to respond to it. We found that recent events had done some of the preparatory work for us: for months rainfall had been well below average, emptying reservoirs, and in their wake had come serious political tensions. Some scientists believed the unusually dry conditions were in part down to the loss of the Atlantic coastal rainforests, but what seemed a more immediate factor was the effect of bulldozers, chainsaws and fire in that arc of deforestation that had spread across the Amazon rainforests. Although the deforestation was taking place thousands of kilometers away, it seemed to be linked with what was happening in São Paulo.

Deforestation to recession?

Considering what we know about the interaction between rainforests and rain, it should be no surprise that deforestation can elevate the risk of drought. What was remarkable in the Amazon was that evidence showed much of the water vapor entering the rainforest from the Atlantic Ocean was not leaving again via rivers, but departing by some other route. Calculations showed the level of evaporation to be bigger than the river flow, and researchers looking at the chemical signature of water falling in the south of Brazil, thousands of kilometers away, discovered the origin of rain there was continental rather than oceanic. 'Sky rivers' provide an explanation as to how these and other observations add up: the long-distance transport of water vapor feeds cloud formation and rainfall in a region that might otherwise be expected to be much drier—and across an area that generates some 70 percent of South America's GDP.

The most affected area was the Rio de la Plata basin, which spreads across southern Brazil and parts of Argentina, Uruguay, Paraguay and Bolivia. It is one of the world's most productive agricultural regions, with soya, maize and wheat cultivation on a vast scale. On arrival in São Paulo, I heard about how the drought had hit production of these staples and hammered dents into the coffee and sugar industries, leading to worker layoffs, reduced profits and higher global prices. Brazil's power sector, too, was in crisis, as hydroelectric dams powered from rains generated by the Amazon rainforests lost significant output. With more than two thirds of Brazil's electricity coming from hydroelectricity, the drought was affecting the whole Brazilian economy. As the drought deepened, cities were subject to power cuts, cutting off power to homes and factories, the air conditioning needed to cool offices in soaring temperatures, and cutting off Internet connections. Brazil had to import power from Argentina and fire up fossil-powered generation capacity.

The severe and prolonged drought was one factor that in 2015 tipped the Brazilian economy into recession. The country fell from seventh to ninth place in the world economic ranking, with unemployment forecast to grow from 3 percent to more

A São Paulo reservoir during the 2014 drought (Getty)

than 10 percent by 2020. The Brazilian drought also had knock-on effects for public health. As the dry period that hit Brazil's southeast got worse, so people caught and stored what little rainwater fell in containers for watering gardens and washing cars. Unlike ponds, which have fish and other predators that eat larvae, these containers made perfect breeding havens for mosquitoes, including those that spread the Zika virus.

Most alarming of all were the predictions of veteran climatologist Antonio Donanto Nobre of Brazil's National Institute for Space Research. Nobre believes that if deforestation in the Amazon is not halted—and if it reaches a point where 40 percent of the Amazon region is deforested—it could lead to an abrupt large-scale shift from rainforests to grasslands. He warns that this could lead to São Paulo drying up, and to an alteration of global weather patterns, including farmlands lying far to the north of South and Central America's tropical rainforests.

South Dakota: farmlands watered by rainforests

The Great Plains of North America, running through the central US north into southern Canada, between the Rockies and Appalachian mountain chains, is one of the world's most productive farming regions. Glacial landscapes extend south to the modern-day course of the Missouri river, marking a time when there were fewer tropical rainforests. About 17,000 years ago the ice began to retreat and was succeeded by prairie grasslands, much of which has today been replaced by agricultural plains of corn (maize), wheat and soya. Vast fields, often 10 square kilometers or more, produce crops on a truly epic scale.

I visited at harvest time. Next to a main road was a column of huge tracked harvesting machines, lined up for their final annual assault on the crops planted the previous May. Laid out in laser-straight rows, the tall crop plants strode across the landscape for miles in all directions. Industrial farming methods were cranked up to an extreme level, including genetically modified varieties of corn and soya designed to withstand herbicides that killed all other plants (or at least that was the case before the recent evolution of herbicide-resistant 'superweeds'). The flat countryside was dotted with the tall cylindrical shapes of silos, where farmers, often operating 2,000 square kilometers, would take their vast hauls of grain. From these monuments of industrial farming the fruits of the season would be shipped to countless businesses fueled by corn: beef producers fattening cattle before slaughter, soft drinks manufacturers using corn syrup to sweeten their fizz, corn snack brands, producers of plant-based plastics and refiners of liquid biofuels among them.

It seemed to me that technology had been harnessed to the limit. That was until I spoke to a farmer who told me about trials of driverless tractors and combine harvesters. These vehicles could plant and reap crops by being driven from a computer on a farmhouse kitchen table or even a city thousands of miles away, as could drones distributing aerial sprays. Yet no matter how sophisticated the convergence of software, hardware, plant breeding, Internet, materials science and satellites, the farming was nowhere without nature—and, specifically, rain.

Located at the very center of North America, the South Dakota corn-belt is highly sensitive to rainfall. So are the banks that finance the farmers who grow the corn and other crops there. A key dividing line is that of the 100° line of longitude. In the past bankers wouldn't loan to farmers plowing the prairies west of there because of the high risk of drought. Further east the risk was lower though yields were still closely related to how much it rains. In drought years farmers can be hit hard, losing money if they don't have government-subsidized insurance.

Although it's very far away from the rainforests of South and Central America, even this distant temperate breadbasket might be affected by the loss of rainforests a continent away, and indeed further away still. The landmark paper by Deborah Lawrence and Karen Vandecar that I mentioned in Chapter 1 sets out why. These researchers modeled the possible impact of forest clearance on long-distance water transfer and found that deforestation in South America, Southeast Asia and Africa could alter growing conditions in agricultural areas as far away as the temperate zones of the USA, Europe and China—including some of the world's most important agricultural zones.

Part of an immense cornfield in South Dakota, USA (TJ)

On a dead straight road in that South Dakota grain landscape, west of the town of Aberdeen, I paused by a new installation of huge silos and tanks connected to gantries and pipes. This cutting-edge piece of infrastructure was linked to a railhead that would move millions of tonnes of material each year. A train with about 120 hopper trucks carrying grain was being unloaded, while another of tanker trucks queued to await its load of corn ethanol, a plant-derived alternative to petrol that would be used to power the huge five-liter engine trucks that charged up and down that highway.

But dependence on rain can impact even the most high-tech infrastructure and poses major financial risks for its investors. People paying into the pension schemes from which investment managers are allocating capital on their behalf might well see corn ethanol as a good financial bet. They might be right about that in narrow financial terms now (although it's probably not a good use for America's grainfields), but if water transfer is interrupted by deforestation they may find their pensions in jeopardy.

A more existential risk is posed to communities where deforestation leaves them more vulnerable to the direct physical effects of extreme weather.

The green shields: how rainforests protect people from extreme weather

In the hurricane season of 2017 several huge storms hit the USA and Caribbean. News reports about storms Harvey, Irma, Jose and Maria pointed out how the run of extreme storms was linked with high surface temperatures on the ocean, causing more water to evaporate, powering devastating winds and floods. A year before, another major storm had hit communities in the Caribbean, but as it affected a very poor country it got less attention. It was Hurricane Matthew.

Hurricane Matthew began to form off the north coast of South America in late September 2016. Beginning as a tropical storm, it grew in strength and a few days later became a powerful hurricane—a Category 4 storm—and on 4 October slammed into the island of Hispaniola, bringing a meter of rainfall and winds of 230 kilometers per hour.

Hispaniola is divided between two countries: the Dominican Republic on the eastern side and Haiti in the west, where the storm made a direct hit. Haiti, one of the world's poorest

Tropical Storm Matthew shortly after it formed on 28 September 2016 over the Lesser Antilles in the southern Caribbean (NASA)

countries, was peculiarly vulnerable to the effects of extreme weather events like this one due to a combination of poverty and massive environmental degradation, particularly deforestation. Hispaniola was once richly forested, but modern aerial maps show an island of stark divisions: the Dominican Republic retains modest tree cover, while Haiti is virtually bare. Most of its trees were first cut for timber and then planted out for coffee, while in recent decades any remaining cover has been used for charcoal or firewood.

With hillsides cleared, the farmland soil that once clothed the land had also been largely destroyed, which not only affected agriculture but caused watercourses to become clogged with sediments, exacerbating the catastrophic flooding that periodically occurs in the lowlands. And it wasn't only in the uplands where the removal of forest cover left the country naked in the face of extreme weather. The coastal mangroves that once fringed much of the Haitian coast had over time become similarly diminished, cut down for fuel once the forests had disappeared on land. Mangroves, especially where they form a broad coastal belt, can slow the flow of water and reduce wave height. In so doing they can reduce the damage caused by storm surges that accompany the intense winds and low atmospheric pressure characteristic of tropical cyclones like Matthew.

With the loss of both the rainforests on land and mangroves by the coast, the violent storm made an unopposed assault, delivering powerful blows from the hills and the sea at the same time. The intense inundation that fell across the uplands quickly turned into gravity-powered water torrents that tore away rocks, mud and other debris that crashed through the valleys and toward the lowlands, smashing bridges, sweeping roads away, flattening crops and flooding homes.

As the storm approached, despite warnings to go to higher ground, many Haitians stayed in their homes. Some were crushed as their houses collapsed onto them. Others were hit by the floodwaters and mudslides that cascaded from the hills, while others fell victim to the surge that welled up in the front of the storm and inundated coastal lowlands. More than a thousand people lost their lives. In addition to the terrible death toll the storm wrecked the economy of an already impoverished country. Tens of thousands were made homeless, crops destroyed and the hillsides further denuded of soil. We'll never know what precise role was played by deforestation in exacerbating these effects, or what difference might have been made had the mangroves and rainforests been in better condition, but the loss of those systems clearly made communities more vulnerable.

Wildlife wipeout

As the rainforests of the New World have been cleared to make way for farming, both by poor farmers and large-scale agribusinesses, the cost is multiple: damage to the water cycle, greenhouse gas emissions and vulnerability to climate change impacts, and also the ongoing depletion of wildlife species.

Take the tropical rainforests of Hispaniola, which once supported many unique species. These included single-island endemic creatures such as the Hispaniola monkey, the Hispaniola edible rat and a species of woodcock that disappeared shortly after the island's European discovery. Other Caribbean islands fared similarly badly. Cuba lost its beautiful species of macaw and a kind of flightless crane that was found there and nowhere else. Jamaica possibly had a species of red macaw but it disappeared, as did its unique species of monkey and ibis. These eye-catching mammals and birds were the tip of an iceberg, however. Departing with them were less obvious

The Jamaican red macaw (*Ara gossei*) may have been endemic to Jamaica but disappeared before being properly described. This nineteenth-century watercolor was by Lionel Walter Rothschild.

creatures, contributing in our century to what is building into a mass extinction of wildlife, driven in large part by the loss of the tropical forests.

The Caribbean islands comprise one of the biodiversity 'hotspots' discussed in Chapter 4—areas where biological diversity is very high and under pressure. All across the tropical rainforests belt of the New World a similar story prevails— through Central America, the forests of the Andes and the coastal forests of southeastern Brazil. These tropical rainforests were (and in part remain) among the most biologically diverse on Earth. The destruction of much of the rainforests on the Pacific slope of the Andes in Ecuador, for example, has already led to one of the largest man-made mass extinctions—including species that were lost before we had time to even give them a name, let alone explore if they might play a role in sustaining our ecological future. In that case, and many others where natural diversity is being depleted most rapidly, it was the loss of forests that was the main driver of decline.

It is not only forest loss and degradation that cause wildlife populations to shrink, but also the fragmentation of the forests. As noted earlier, the tropical forests have been shattered into tens of millions of separate pieces and, as a general rule, the smaller an area of forest the fewer species it will contain. Modern conservation ecology has established that even if a forest fragment has all of the species found in the block from which it has splintered, the number will decline to reach an equilibrium based on the size of the fragment. This is a vital fact that must be reflected in how we plan to avert the potential extinction of the legions of species now at risk of being lost, including some of the world's most beautiful and charismatic animals. They include the giant otter, pink river dolphin (the rainforest river habitat of which is being fragmented by large dams), the South American tapir and the curious human-looking red-faced Uakari monkey, which carries the name of an already extinct tribe.

The idea that fewer species might be expected to be found in smaller patches of habitat was first put forward in 1967 by Harvard Professor Edward Wilson and his colleague Robert MacArthur in their 'theory of island biogeography.' One example that they drew upon was Barro Colorado, an actual island that was created in 1913, when the Chagres river was dammed in the construction of the Panama Canal. When the waters rose, an expanse of tropical rainforest was flooded, leaving only hilltops dry—including what is now the 15.6 square kilometers of Barro Colorado. The new island made the perfect real-world laboratory to test a theory that predicted that the number of species there would decline over time.

I met Wilson in 2016. Although he was then 87 years old he lacked none of the clarity that had earned him a worldwide reputation as an ecologist. He explained how he'd been able to use data collected from Barro Colorado by 'some very good students of birds who made complete censuses before and afterwards.' The data confirmed his and MacArthur's theory of biogeography, and he 'ran a series of experiments in the Florida Keys on these small mangrove islands and got the effect again.'

Barro Colorado was a real island surrounded by water. Many areas of tropical forest today are virtual ones, surrounded by plantations of trees destined for pulp mills, fields of cocoa trees, oil palms, smallholders' farms and soya and rubber plantations. But the island effect is the same: species decline, and local and then global extinctions can follow. There are a number of reasons why this effect occurs. One is how conditions change near to forest edges—for example, where it becomes less humid. We've seen how this can affect trees and thus carbon emissions. It can also affect other plants and also animals such as amphibians, with knock-on impacts, for example, for nectar feeders and creatures that prey on frogs and salamanders. The smaller the area of forest, the greater the effect of new edges will be. The tree composition changes too, with more generalist species living near the edges and deep forest trees tending to die out. As the tree species mix alters, so does the mix of animals that depends upon them, such as fruit and seed-eating species.

Then there is the effect of pathogens hitting small isolated populations. Should a disease reach an isolated group of animals or plants, then that local population might be wiped out. If disease repeatedly hits a species with limited ability to move between isolated fragments of suitable habitat, then each sub-population that is lost will over time add up to total extinction. One-off events such as hurricanes and volcanic eruptions can also take a deadly toll on tiny isolated populations.

On top of this are the insidious effects of climate change. As the world warms and rainfall patterns and seasonal cycles alter, so many animals and plants will need to move to new places. For example, species living in tropical cloud forests will need to move to new habitats at higher altitudes that retain the cool temperature and moisture regime previously found in forests lower down. As the warming progresses, there will eventually be no higher places left to go, and in the meantime reduced options for getting there. Extinctions will follow.

These effects are especially important for animals that need a large area of forest, without which they cannot maintain a genetically viable population. As a general rule it is believed that if animals are not to suffer the effects of inbreeding, at least 500 or so individuals need to be in a population. For large predators like jaguars that live at low density, this means that for their effective long-term conservation very large areas of connected habitat are required. A few animals left in isolated forest 'islands' will simply not be enough.

Wilson and MacArthur are among a group of conservation scientists who over the years have highlighted the profound effects of fragmentation and isolation, showing how these trends pose a fundamental threat to large numbers of species. Wilson told me how the effects of fragmentation had to be urgently reversed. 'Island biogeography is now a stanchion of conservation science and it is why we need to protect the biggest areas we can get. It's the only way we're going to save the great majority of species.' Given the particularly high diversity of species found in hotspots, then the application of this rule in those places emerges as a clear conservation priority.

As elsewhere in tropical rainforest, there is a further set of pressures compounding the effects of fragmentation: hunting and poaching. This causes more than the decline and loss of individual species, as the loss of one species generally changes conditions for others.

A somewhat encouraging piece of research, however (also conducted on Barro Colorado Island), suggests that some forests have an inbuilt resilience against progressive species loss, with their original high level of bird and mammal diversity providing some insurance against the loss of ecologically important animals. For example, many large trees found in the South and Central American forests appear to be adapted for dispersal by large animals, including those hunted to extinction by people several thousand years ago. One study in a Central American rainforest used miniature radio transmitters to track the movement of agoutis, rodents that look a bit like long-legged guinea pigs. The animals were found to be constantly moving seeds between cache sites, as they stole, moved and re-hid each other's stores. Rather than eating what they took, the tendency was to re-hide them. It was estimated that individual seeds were moved up to 36 times with agoutis shifting more than a third of the items they got hold of more than 100 meters. Video material revealed how the rodents repeatedly re-cached each other's stores, meaning that particular seeds could travel a long way. An estimated 14 percent of seeds survived to the following year, at which point a new crop became available, leaving those from the previous year a better chance of not being eaten and instead growing a new tree.

This piece of research demonstrated how communities of rodents can be highly effective long-distance seed dispersers and could explain why large-seeded trees that evolved alongside long-lost large mammals have not yet disappeared. Agoutis are, however, a favored food animal that falls prey not only to predators such as the jaguar and ocelot, but also human hunters. With this kind of research in mind we might see even more good reason to protect the creatures that remain in what are, from the point of view of their original wildlife, severely depleted forests.

8 PROTEST AND SURVIVE

As awareness grew about rainforest destruction, campaigns began to make some impact in protecting what was left

The mix of factors that lead to deforestation, declining forest health and loss of wildlife are often quite complex—but making a difference to public opinion, and impacting on politicians and businesses, is about conveying simple messages. So it was that, one spring morning in 1993, I stood outside Harrods with Friends of the Earth colleagues. Each of us held a placard with a single letter, spelling out M-A-H-O-G-A-N-Y I-S M-U-R-D-E-R D-O-N-T B-U-Y I-T.

Press photographers snapped away, TV cameras rolled and we did interviews putting rainforest protection on the political agenda. It was a step in a campaign aimed at ending the import of illegally harvested mahogany to the UK. We had published research, and sent it to importers and retailers, and had now taken to Britain's high streets. The idea was to put market pressure on a trade that was devastating rainforests, and to highlight the whole issue of logging and deforestation in the rainforests.

Mahogany

With its tight straight grain and rich-reddish polished sheen, mahogany had for centuries been regarded as the height of

Friends of the Earth protests outside Harrods in May 1993
(Friends of the Earth)

quality and fashion. Several species of rainforest tree produce timber that has been called 'mahogany,' but the genuine article is derived from a species called big-leaf mahogany (*Swietenia macrophylla*), whose natural range extends through New World rainforests from southern Mexico and to Peru and Bolivia, as well as the Amazon rainforests of Brazil.

The trunks of old trees were the most sought after as the wide boards needed to make large tables, cabinets and paneling could be cut from their huge stems. From the early years of European colonization, mahogany was used for these purposes as well as for shipbuilding and construction. Philip II of Spain used mahogany for the interior of the Escorial Palace, the construction of which was begun a few years before his Armada of timber ships sailed on England.

Early demand was fed from a closely related species of mahogany tree found in the rainforests of the Caribbean islands. But as supplies of that wood from Cuba, Hispaniola and Jamaica were exhausted, so logging activities shifted toward Central and South America. As might be expected, availability of the most valuable large-stemmed trees of this slow-growing and long-lived species began to decline. By the the time we lined up outside Harrods, most of the best-quality specimens were to be found inside protected areas—National Parks and indigenous reserves. Taking such trees was illegal but didn't deter those looking for profit, especially as corruption of public officials could drastically reduce the risk of prosecution. And it wasn't a small-scale trade. Brazilian government figures suggested that between 1982 and 1992 about 2 million cubic meters had been illegally cut in indigenous reserves, with more than two thirds of that exported. Its destination were high-end consumer outlets, many of them in Britain and the US, where tropical hardwood furniture was a thriving and profitable business.

As we've seen, logging, whether legal or illegal, is often the first step in a process that leads to total deforestation. An analysis (published some years after this campaign) showed how across a 203-million-hectare area of the Brazilian Amazon some 16 percent of selectively logged areas were completely deforested within one year, and after four years nearly a third of the logged forest had completely gone. Logging is never just about removing a few prize specimens.

The Friends of the Earth campaign emerged from a meeting we had with the Amazon Working Group—a coalition of environmental campaigners, social justice advocates, indigenous and rubber tapper organizations. They were pressing for policies to protect the rainforest and its people and wanted to take their cause to the global level. They told us of violent—and often fatal—attacks against indigenous people who had tried to resist the loggers. Previously isolated communities had also been devastated by disease outbreaks, including influenza and measles, just like their ancestors centuries before. Raising awareness about the killing of the forest people was to be at the core of the new campaign and the reason why it was called 'Mahogany is Murder.' Its aim was to spread a boycott of the timber so as to close down demand in Britain and hopefully beyond.

We launched the campaign in late 1992 with the publication of a short report called *Mahogany is Murder* by our fellow campaigner George Monbiot. He had recently published a groundbreaking book called *Amazon Watershed*—a first-hand account of what was going on in the forests. George had exposed how a vast military project risked establishing a new front of deforestation in the north of the Amazon and how it was the timber industry that posed a growing threat to the forests, rather than the ranching of cattle which was back then getting a lot of attention. Crucially, he had traced timber illegally cut in Indian reserves all the way back to retailers in Britain. We added our own reports and evidence, including official Brazilian government documents and testimonies from indigenous people that I had procured on visits to Brazil.

An opening call to action came in a letter to the British people from the former Brazilian environment minister José Lutzenberger. He pointed out how the trade was largely based on illegally sourced timber and that most was coming from indigenous reserves. He called on Britain to take action by not buying it. Such a message from a former member of the Brazilian government carried weight; so did the timing with our campaign launched during the year that marked the 500th anniversary of Columbus's arrival in the New World. The fact that half a millennium later the same genocidal consequences still accompanied the process of resource extraction provided a powerful context.

We didn't have a big budget to promote the campaign but secured cheap roadside poster space through some friendly media buyers and hired prominent sites along main roads in and out of London to proclaim our slogan. We also got the message out via a short film written by McCann Erikson, one of the leading London advertising agencies, which featured blood spilling out over a mahogany toilet seat as a voice-over spoke of the devastation being caused by the trade. The ad was shown in independent cinemas until complaints from the timber industry and Brazilian Embassy in London led to it being banned—an additional spur to publicity. We put it on a new medium called the Internet (where it can still be seen on YouTube).

Perhaps more edgy than a film that risked offending some people's sense of decency was the tactic of 'ethical shoplifting.' This involved campaigners entering high-street shops, removing mahogany furniture and taking it to the local police station, where we would say we had reason to believe the timber had been illegally extracted from indigenous reserves in the Amazon. While the police couldn't do very much, the message spread, as TV crews, reporters and radio broadcasters came along.

We also enjoyed the communications benefits from a treasured campaign asset—a huge inflatable chainsaw, 12 meters long, with a great floppy blade. It was an impressive bit of kit. Campaigns Director Andrew Lees had met a couple of people in a pub in Camden who made inflatable props for rock bands, including the pink pig that was famously floated by Battersea Power Station for the cover of the Pink Floyd album *Animals*. In due course they showed up at our Shoreditch headquarters with our unlikely new communications device and showed us how to

Not a welcome sight for your local DIY store manager
(Friends of the Earth)

inflate it, with a pump that could be revved up to sound like an oversized chainsaw. They said it could be mounted on a van and driven around, and if we wanted they'd help us do that.

So, two weeks later the inflatable chainsaw had its first of many outings, to a home improvement store in North London that belonged to the Wickes chain. Like other DIY stores they sold a lot of wood and we'd decided to target them in calling for measures to ensure that only legal and sustainable timber was sold in the UK. The Air Artists arrived, blew up the chainsaw, and set off around the car park, to the amazement of the media crews we'd invited. Emblazoned along its length were the words 'Stop the Chainsaw Massacre' and the shot on the evening news was of the saw pursuing the store manager.

The companies on the receiving end of this kind of attention were appalled. At first they employed PR companies to put out press releases saying that selling tropical rainforest timber was legal and that they took every care to ensure that what they sold was responsibly sourced. The fact that they could not show that either of these things was true, and that we had plenty of evidence to demonstrate the contrary, meant we campaigned on. And it worked. Several high-profile store chains, including B&Q, Sainsbury's and Great Mills, announced that they'd stop stocking Brazilian mahogany, citing the impact of the trade on indigenous rainforest peoples as the reason. The campaign aimed to add fuel to parallel attempts by WWF to get a new timber certification scheme up and running. When, later that year, agreement was reached to establish the Forest Stewardship Council (FSC), creating chains of evidence from forests to shops to prove that wood taken from forests was legal and sustainable, the pressure on the timber trade was elevated still further. After all, if certified sustainable timber was available for sale, why would anyone stock anything else?

All this began to make a difference, with serious ramifications for the timber supply companies. On mahogany we kept up the pressure in the UK and tried to internationalize the campaign. Even with new timber certification schemes, however, we knew that British consumers and the companies who sold them mahogany products would not, on their own, be able to solve the problem. Wider official action was needed. So we decided to press for getting big-leaf mahogany to be listed as a threatened species under the Convention on the International Trade in Endangered Species (CITES).

CITES

The CITES treaty is intended to reduce the pressures on animal and plant species arising from international trade. In the 1990s, elephant ivory, whale meat, rhino horn and tiger skins were more familiar agenda items at CITES meetings, but plants, including trees, could also be listed (and indeed the biggest single group of controlled species listed in the treaty is tropical orchids). Our idea was to get mahogany included so that any of it entering international trade would have to be accompanied by official permits to confirm that it had been legally and sustainably sourced. Any mahogany without such a permit could be seized. This would be a major step forward in securing indigenous reserves and other protected forests.

It was decided to target the next meeting of CITES that was to be held, at Fort Lauderdale, Florida, in November 1994. We knew media coverage of the issues was essential to ramp up political pressure on the government ministers attending, and we decided to focus on a timber ship arriving at Heysham in Lancashire from Belém, a major port at the mouth of the Amazon, in Brazil. The ship was called the *MV Wind* and was carrying more than 2,000 tonnes of mahogany. My colleagues were waiting. On the dockside we had installed a massive PA system, through which we bombarded the ship with the sounds of rainforest birds, animals and insects, interspersed with that of chainsaws and falling trees, and unfurled a 70-foot banner that once again proclaimed 'Mahogany is Murder.'

At that point I was already en route to Fort Lauderdale, carrying, in addition to a range of research reports and press materials, 3,000 cardboard drinks coasters. On one side they bore the circular campaign logo and the message 'Mahogany is Murder'; on the other, 'CITES listing or consumer boycott?' At international meetings like this much business is transacted informally in the bars and hotels around the conference center. My plan was that the coasters could show government negotiators that they had a choice: either introduce an orderly set of controls to protect the forests and their peoples

A giant mahogany tree growing deep in an indigenous reserve in the Peruvian Amazon (TJ)

from illegal timber cutting, or face the chaos of a consumer backlash. It was a peculiarly effective, low-cost bit of media, with people at the conference talking about the proposal to list mahogany sponsored by the governments of Costa Rica and the Netherlands. Timber industry representatives who'd turned up to fend off the proposal were highly agitated. I slogged it out with them in media interviews, including on the subject of whether a voluntary 'gentlemen's' agreement would be sufficient to halt the impacts on the indigenous people and the forests.

The proposal to get mahogany listed on CITES didn't work at that meeting, mainly due to its rejection by Brazil and Bolivia, the two main countries exporting the wood to global markets. But, while that marked a setback, taking the campaign transatlantic had positive benefits. One was seen the following year when our friends at the San Francisco–based Rainforest Action Network joined the fray. Activists led by director Randy Hayes took to boarding ships to prevent mahogany cargoes being unloaded. They also went to American DIY stores, including Home Depot, and pressed for the kind of boycott we'd sought in Britain. Greenpeace, who had a newly established Brazilian office, also joined the campaign.

As the pressure continued, so the Brazilian government had to react—not least because their trade was being impacted. More than 35,000 cubic meters of Brazilian mahogany had come into the UK in 1993, but by 1996 that had dropped to less than 11,000. And countries generally try to avoid being embarrassed internationally. It's not good for business or investment. So when questionable practices like the invasion of indigenous lands by loggers cause campaigns to break out in other countries, it pays to do something about it. Once political will to act had been galvanized in Brazil, the country's own official environmental protection and indigenous bodies launched investigations that found only about a quarter of mahogany shipments were legal. A crackdown followed. Timber consignments were seized; companies had their timber felling licenses revoked; and a moratorium on new logging licenses was introduced.

Action against illegal logging became an established Brazilian government policy and mahogany finally got listed in CITES in 2002. This official action to curb illegal logging, driven in part by campaigning pressure, really made a difference. But, even as this progress was made, the depredations of the timber trade were overtaken by even more serious and devastating pressures—the total conversion of forests to agriculture.

Eating the forests: the Soya Rush

While during the 1980s and 1990s most campaigning effort for the South American rainforest concerned logging and cattle ranching, it became clear by the 2000s that it was expanding demand for a range of other commodities that had become an ever more potent and important threat. During that decade the rate of forest clearance was rising steeply to make way for oil palms, beef cattle pastures, pulpwood plantations and, in particular, fields of soya beans. When it came to soya, the main frontier of deforestation was in the vast basin of the Amazon. During 2004 alone, some 12,000 square kilometers of land within the Brazilian Amazon was planted with the crop.

The rapid expansion of soya was enabled by new infrastructure. In 2003 the global commodities giant Cargill opened a $20 million facility at the port of Santarém on the confluence of the Tapajós and Amazon rivers. Then, in 2004, Brazil's Federal Government announced plans to pave the entire 1,700-kilometer route of the BR163 highway, allowing year-round access from Cargill's soya port into the heart of the western Amazon in Mato Grosso. The road soon became known as the soya highway.

Cargill's soya export dock at the confluence of the Tapajos and
Amazon rivers (Greenpeace)

Expanding soya cultivation was a major reason why 2004 saw
the second highest level of deforestation ever recorded across
the Amazon basin rainforests. Brazilian soya was (and remains)
destined mainly for animal feed, driven by fast-rising global
demand for meat and dairy foods, notably in China. As soya
prices went up, land rose in value, and soya cultivation was
moved into new areas across the southern and eastern fringes of
the Brazilian Amazon states of Para, Mato Grosso and Rondônia.
It caused, in effect, a 'Soya Rush' to take hold.

As the scale of the mounting threat to the forests became
apparent, so campaigners—and Greenpeace in particular—
began to take an interest, with activists seeking to block soya

from being shipped out at the new Santarém soya port facilities. John Sauven, who became Director of Greenpeace UK in 2008, was one of the architects of their rainforest campaign. 'Cargill acted as a kind of magnet for encouraging farmers to come in and grow soya,' he told me. 'They had the port and facilities for receiving soya beans and exporting them. They also put a lot of capital into farmers. They paid them for the seed and equipment and everything else. Then they bought the product.'

Cargill and the other big trading houses—ADM, Bunge and Dreyfus—were the hub from which the soya flowed from the fields to the global market and were the obvious target for action. These vast, but relatively anonymous, commodity companies would not easily be swayed by Greenpeace, however. Neither would the soya growers who were actually cutting the forests.

'There was absolutely no way we could touch the farmers directly,' Sauven explained. 'If we ever tried to do anything or blockade anything or go onto any land connected to them, they'd just shoot us. But the funnel that soya had to go through to get onto the open market was bizarrely small. There were five companies and they were taking most of the soya to global markets. But those too were pretty much untouchable, as most were privately owned.'

With no shareholders to influence and no real high street or brand presence to target, Greenpeace looked for other pressure points. The campaigners went in search of links to big consumer-facing brands who would be aware that if they wished to do good business in Western markets they'd have to do it without deforestation as a consequence. So Greenpeace set out to discover to whom Cargill was selling the soya. They identified shipments from deforested areas going to Santarém and from there monitored ships leaving the port, especially for shipments heading toward the UK. When the ships arrived at British ports, researchers watched soya shipments being unloaded and waited by roadsides for the trucks to pass by. 'It was quite laborious,' recalled Sauven. 'Waiting at a lay-by for lorries to leave the plant and follow them up to some place in the middle of nowhere in Scotland, or the northeast. We kept on following them.'

They found that the soya was being sent to a central plant, crushed into animal feed and then trucked to hundreds of mostly small farms. 'Eventually,' Sauven said, 'we followed one of these lorries and it went into a massive, Cargill-owned chicken processing factory in Herefordshire. Now that was a lucky break because now we had a big chicken producer owned by Cargill.' One of the Greenpeace team pretended to be a teacher and got an appointment to see if it would be possible to bring his pupils in from the local school. The chicken farmers showed him round, and everywhere he went he saw posters with McDonald's logos. The farmers proudly explained that they supplied the fast-food chain with its chicken nuggets.

That was it. Greenpeace had established a full set of links: from deforestation in the Amazon, to soya arriving at Cargill's processing plant in Santarém, from there to Liverpool, where the soya was crushed, then to the chicken factory in Herefordshire producing nuggets for McDonald's. The next task was to make public the connections between forest loss and McDonald's, and in so doing to encourage that major company to insist that deforestation was eliminated from any soya used in its products.

Greenpeace released a report called *Eating up the Amazon*, and volunteers dressed up as seven-foot-tall chickens and picketed McDonald's restaurants. 'It caused a huge storm,' said Sauven. 'There was massive amounts of press coverage about how McDonald's was destroying the Amazon. They withstood the heat for about twenty-four hours before saying how it was outrageous that Cargill had never told them. They said they

Ecochickens put the spotlight on McDonald's (Greenpeace)

were shocked and wanted nothing to do with destruction of the rainforests.' After that McDonald's basically joined the Greenpeace campaign. The world's largest burger franchise insisted that if it was to continue doing business with Cargill, then deforestation must end immediately. The campaign group and biggest fast-food brand comprised an unlikely alliance, but action followed.

In July 2006 a moratorium on deforestation was announced, whereby none of the big traders would buy soya from any farmer involved in deforestation. Two of Brazil's farm industry

Forest pump Water not only runs through rainforest rivers but also flows through leaves and into the air, generating rainfall thousands of miles away. Danum Valley Conservation Area, Sabah, Borneo, Malaysia (above); Kaieteur Falls, Potaro River, Kaieteur National Park, Guyana (below)

Ponds in the treetops Bromeliads catch water and provide drinking and breeding opportunities for a range of animals. Cave of the Oilbirds, Colombia.

Land of the frogs Clockwise from top left: long-nosed horned frog (*Megophrys nasuta*), Borneo; tiger's tree frog (*Hyloscirtus tigrinus*); harlequin poison frog (*Oophaga histrionica*), Colombia; red-eyed tree frog (*Agalychnis callidryas*), Costa Rica; another Colombian harlequin; barred leaf frog (*Phyllomedusa tomopterna*), Bolivia.

Bug life Insect diversity is extreme in rainforest environments. Clockwise from top left: giant saturnid moth, Ecuador; arctiid moth, Costa Rica; lucanid stag beetle, Peru; tettigonid katydid, Peru; psychedelic grasshopper (*Phymateus saxosus*), Madagascar; cerambycid longhorn beetle, Madagascar.

Global cooling Giant trees like this one hold a high proportion of the aboveground carbon trapped in the

Circular economy Fungi and millipedes are among the myriad species that assist in the rapid recycling of nutrients in rainforests. Cup fungi, Choco rainforest, Colombia (above); giant millipede (*diplopoda*), Northern Madagascar (below).

Wonder The sheer strangeness and unfamiliarity of rainforest flora and fauna can be dazzling. Lowland streaked tenrec (*Hemicentetes semispinosus*), Madagascar (above); comet moth (*Argema mittrei*), Madagascar (below).

Floral tapestry The moist, still conditions that prevail in tropical rainforests require flowering plants to invest in blooms and nectar to attract animals to move their pollen. Passion Flower (*Passiflora vitifolia*), Costa Rica (above); orchid, Borneo (below).

associations, who between them controlled 92 percent of the country's soya production, signed up to an agreement not to supply soya produced on land deforested after that point. A working group comprised of the trading companies, campaign groups, global brands, the Brazilian government and the Banco do Brasil (the main source of credit for soya farmers in the Amazon) was formed to oversee the accord.

Aerial surveillance and satellite technology was used to monitor the 76 municipalities where nearly all Brazilian Amazon soya was produced. Any farmer found to be in contravention of the forest clearance moratorium was to have contracts with traders like Cargill canceled. First time offenders were allowed to let the forest grow back, but repeat offenders were not only barred from trading but also from securing loans from the Banco do Brasil. The moratorium was initially adopted for two years, but was repeatedly reinstated and remained in place a decade later.

One interesting aspect of this change in attitude towards the Amazon is that in the years that followed the adoption of the moratorium Brazil still managed to increase soya exports. It did so by using new agricultural technologies, better practices to conserve soils and former pastures were cleared long before the moratorium came into effect.

Similarly effective pressure was subsequently applied to the beef industry. As old pastures were being used to grow soya, so the clearance of new ones had been taking place to make way for more cattle to meet growing global demand for beef and leather. This posed a new fast-growing threat to the Amazon rainforests and Greenpeace again sought connections between deforestation and products in supermarkets and leather seats in luxury cars. As was the case with soya, beef was going to global markets via a tight bottleneck of major trading companies. Sauven told me how his team discovered that, in relation to beef, it was an even smaller number. 'It was incredible. Whereas we'd found soya coming out through five companies, when we looked at cattle it was three—just three companies deciding whether much of the Amazon rainforests were going to be destroyed or protected.' Global brands exposed to the risk of being linked with rainforest destruction mobilized on this issue, too, with an agreement forged in October 2009 to ensure cattle brought for slaughter would be traced back to individual farms, at which point it would be verified that no recent deforestation had taken place.

By 2012 the campaigns of the 1990s on illegal logging and the mahogany trade, followed by those during the 2000s for sustainable agriculture, had made a very considerable difference. In that year the deforestation rate in the Brazilian Amazon had dropped to 4,571 square kilometers—an 80 percent reduction from the peak level in 2004. It had happened through a combination of consumer action, the mobilization of global brands, commercial and trade pressures, the use of remote sensing technology and, crucially, political will inside Brazil. Even though that political commitment would in the years ahead become considerably weaker, it did reveal that it was possible to create the kind of circumstances under which the rate of forest loss could be dramatically reduced.

Campaigns like these have been vital stimulants for action in Brazil. But there are other models, where it is an appreciation of the practical benefits of rainforest to the national economy that make the difference. The great example of this is the diminutive Central American republic of Costa Rica.

9 FOREST DIVIDENDS IN COSTA RICA

The destruction of the rainforests is not an
inevitable result of economic development—
Costa Rica has shown otherwise

The town of Liverpool lies just inland of Limon, Costa Rica's
main sea port on its Caribbean shore. In common with its English
namesake, it is a commercial hub and on its main drag you
notice hundreds of refrigerated containers, stacked and loaded
with bananas and pineapples. The big global fruit companies
are here in force—Del Monte, Dole and Chiquita among them—
preparing to get produce from fields to distant consumers. It
is a major logistical effort to unite the fruit produced here on
volcanic soils drenched with tropical rains and with strong sun
to distant markets in North America and Europe. Close to the
city fast-growing trees were planted to provide the wood for the
countless wooden pallets needed to ship the fruit to market.

I watched from the quayside as convoys of container trucks
headed towards the loading area and cargo vessels took produce
aboard, while on another wharf tourists filed off a cruise ship.
A lot of fruit was going out and a lot of people were visiting,
with the economy strengthened in both directions. It was
all taking place at the spot where, in 1502, Columbus made
landfall, dubbing this country 'Rich Coast.' And—unusually—
Columbus might still recognize this coast, as forest cover remains

on the flat coastal plain and mountain ranges behind. The rest
of the country, too, retains much of its forest, bucking Central
America's trend toward deforestation. Indeed, since the 1980s,
Costa Rica has actually doubled its forest cover.

This remarkable fact, and the conservation culture that lies
behind it, has roots that go back to the 1850s, when laws were
passed to protect the country's incredible wealth of natural assets,
including wild game for hunting, timber resources and fisheries.
They were enhanced in 1895, when further protection was set
down for water supplies and rangers appointed to regulate the
burning of pastures by cattle ranchers. Much later, in the 1960s,
many National Parks were designated, protecting some of the
country's amazing tropical rainforest. A National Park service
was established, enabling the lines marking the boundaries of
protected areas shown on maps to have practical effect, in the
face of encroachment by loggers, ranchers and miners.

Despite these steps toward conservation, Costa Rica's rate of
deforestation during the 1970s and 1980s was rapid. As in many
of its Central American neighbors, large tracts of forest were
turned over to ranching, mainly to supply cheap beef for the USA.
By the 1980s, nearly a third of the country had been cleared for
cattle pasture, though much of it lay abandoned and worthless.
But in that decade ministers became convinced that the forests
were worth more alive than dead, not least because of the rain
they produced, which ensured farming remained productive and
kept the country's extensive hydropower going. They also realized
that the incredible wildlife diversity of the nation could be a huge
draw for tourism. On all counts they were proved correct.

This shift in perspective followed the establishment during the
early 1980s of a new government department responsible for the
environment and energy. Carlos Manuel Rodriguez was a minister
in that department during the 1980s and told me something of
the discussion that went on in government. He explained how past

A pure river flows from intact tropical rainforest on the Caribbean slope of Costa Rica (TJ)

deforestation in Costa Rica was driven by short-term economic pressures. 'We are quite close to the USA and during the 1970s the fast-food industry was growing rapidly. There was a huge demand for beef and we introduced large-scale ranching to meet it, and the rate of deforestation to make space for it was extremely high. During that decade we had the highest deforestation rate per capita in the world. At that time the market did not recognize the value of nature and the forests.'

Even though National Parks had been set up, that on its own was not enough to conserve the country's forests. Attention turned more toward economic remedies. 'We understood that forest conservation could only be achievable if the value of the standing forest could match the cost of opportunity of not doing cattle ranching. So we put in place a lot of forest conservation incentives, using subsidies, tax breaks and payments. This began to lower the level of deforestation.'

This kind of policy could only continue, however, if there was a clear benefit to the national economy. 'The finance ministry guys couldn't properly measure the value of our investment. We needed to show the economic benefits of protecting nature.' So Rodriguez and his team began to make the case for forest conservation based on raw economic arguments—at first in relation to nature-based tourism. 'Ecotourism was growing fast and we realized we could generate some information as to the value of protected areas.' And then they realized it was possible to quantify the value of the forests in relation to the hydroelectricity industry (which provides most of the country's power). They showed that trees captured and released water to rivers that topped up reservoirs that powered turbines. Less forest meant less water, and that meant less power.

Having previously been told by the finance minister that health, education, new infrastructure and poverty reduction were priorities, and that nature was not, Rodriguez presented his new information. 'When we had this work completed I went back to the finance minister, but this time with some economists. When he saw these guys with me he began to talk to them and they were speaking the same language. This was a turning point, and now the economics of nature is institutionalized in Costa Rica.' Not only were natural areas protected, large tracts of degraded land were restored too. 'In the late 1980s we had only 21 percent of forest cover, but this is now 52 percent.'

Rodriguez told me how this impressive outcome was achieved at the same time as improving living standards. 'When the forest was at its low point the GDP was around $3,600 per person; now that the forest area has more than doubled, the GDP is around $9,000.' And on the energy side, too, there was impressive progress: 'In 1985 Costa Rica generated half of its energy from renewables, and half from fossil fuels. Twenty-five years later we generate 92 percent from renewables.'

One remarkable thing about the journey Costa Rica has been on is how consistent it has been, building on success to embed conservation deeper into the national fabric. In 1988 a national conservation strategy was adopted (the UK doesn't have one, incidentally), a theme that has been reinforced by successive presidents, including José María Figueres, who on his election in 1994 proclaimed, 'We will build a constructive alliance with nature.' On the back of that commitment, so rare from a head of state, Figueres sought to adopt a coherent approach to conservation, combining government efforts with the many private initiatives going on across the country; for example, elevating the protection of private reserves through the same rules and regulations that were applied to official National Parks. The 1996 Forest Law helped to reorient forests away from logging and toward conservation, again in part on the basis of the practical value of doing this, raising awareness about the value of intact forests for carbon fixation, water supply and hydropower, protecting wildlife and their scenic value—and thus their role in sustaining tourism.

While noting these benefits for conservation, it is important to be reminded that all of this was done with national interest very much in mind. As Rodriguez asserts: 'We have realized that there is no long-term economic growth without protecting the health of the ecosystems. This is all about humans, not about nature.' Costa Rica is a modern, vibrant market democracy. The main shopping district in the capital San José could be in Europe or the USA. International companies have made major investments, not least American high-tech firms, attracted not only by a motivated and educated workforce but also by the clean air needed for manufacturing advanced digital products.

The pure atmosphere derives from Costa Rica's success in conserving and expanding its forests, and from making the switch to renewable electricity. Some of that is coming from hydroelectric dams being fed by rivers topped up with rain generated by the rainforests. Those rainforests are also the main attraction for millions of foreign visitors. Indeed, since 1999, tourism has earned more foreign exchange than the combined exports of the country's three main cash crops: bananas, pineapples and coffee. About 2 million tourists visit each year—a huge number, considering the population of the country is only about 4.5 million.

Regenerating rainforest

I visited a former farm called Selva Bananito, on Costa Rica's Caribbean slope, where rainforest is being regenerated following clearance for cattle pastures. It is one of hundreds of degraded landscapes—cattle pasture, banana plantations and logged areas—where forest is being restored. Still essentially a large farm, with some original rainforest on and around it, it now operates as a tourist lodge. The project managers have put in place conditions for natural regeneration and have also been taking seedlings from the rainforest trees and planting them in degraded areas. It is working. In one patch of trees planted less than a decade ago, I saw a range of forest birds, including trogons. Winged fruit eaters like these will bring the seeds of

Selva Bananito, a Costa Rican farm where rainforest is being restored
(TJ)

other trees in their gut and deposit them in their droppings—
one factor that will progressively enrich the species mix in the
plantings and accelerate their transition into new rainforests. In
a few hundred years it might be that ecologists will consider this
to be a primary forest, as we have designated those areas of the
Americas that were ancient cities and where natural rainforest
regenerated. The main difference now, of course, is the fact that
people are deliberately re-establishing the ecosystem.

Staying for a few days at Selva Bananito I saw a very practical
example of how during the years ahead the grand visions of

some conservationists to reconnect fragmented areas of forest might be realized. One ambitious project, the so-called 'Path of the Panther,' seeks to recreate an unbroken forest spanning the length of Central America, stretching out from the International Park that protects the highland forests between Costa Rica and into Panama.

Selva Bananito is thriving as a tourism destination and among the staff working there were young German students volunteering before their university studies. They were passionate about the forests and gave me hope that a new generation might have a positive view on the value of rainforests, unlike the recent decades of destruction. There were also Danish, French, American and fellow British tourists. All had come to stay in the amazing wildlife-rich surroundings and in so doing were not only helping outcompete the economic forces of deforestation, but were a part of the process of reforestation.

At another rainforest reserve, the famous La Selva Biological Station, where so much research into tropical rainforest ecology has been conducted, I met a guide who told me something about the country's recent journey to being a top ecotourism destination. He was in his mid-thirties and explained how at high school, in the late 1990s, he had been able to take a course in ecotourism. The fact such training was an option on the curriculum spoke volumes about the joined-up economic strategy that has been adopted by Costa Rica and which has forest conservation at its heart.

The guide spoke perfect English, had forensic knowledge of forest ecology and, as a result of having mastered the tourism business, was able to make a very good living for himself and his family. He met people from all over the world and was one of thousands of Costa Ricans who worked in the country's growing wildlife sector. I learned how during their schooling many

When the trees come back, so do birds like this trogon, bringing new
tree seeds with them (TM)

Costa Rican children not only had opportunities for formal training in ecological subjects but were also offered volunteering opportunities to work in a National Park or other protected forest, helping to regenerate degraded areas. This deliberate policy to foster awareness and achieve economic benefit from nature is a core element in a 'green growth' strategy that has contributed to what is perhaps the most meaningful indicator of all: national happiness.

Well-being

A recent poll by Gallup found Costa Rica to have the highest level of well-being in the world, as well as some of the longest-lived people: life expectancy is 79.6 years—higher than the US. The country delivers this level of happiness using about a quarter of the resources of typical Western nations. Alongside far-sighted environmental policies, Costa Rica has for decades made huge investments into social programs, enhancing health, education and housing.

Traveling in very remote areas, far from tarmac highways and through villages where poverty is evidently still prevalent, I couldn't help noticing new schools and clinics. Government housing meant that most people had a place to live. I saw modest rural homes set among fruiting and flowering trees, with flocks of parakeets trilling in the branches. I could think of worse places to live, including some other Central American nations, with their starker social divides.

The fact that Costa Rica has been able to do this, in contrast to their neighbors, is down to a number of factors. One of them is how in 1949 this little country adopted the far-sighted and distinguished policy of abolishing its armed forces. At a stroke, massive financial resources were freed up for social progress, while at the same time there was the added benefit of their being no military available for participation in coups or civil wars. Internal conflict has blackened the recent history of Nicaragua, El Salvador and Honduras, in the process causing massive deforestation. Costa Rica has meanwhile been stable, green and inclusive.

Large-scale conservation and green growth programs now getting underway in other tropical rainforest regions won't all be able to emulate Costa Rica in this respect. But as we go forward, its example shows that it's possible to have a modern economy, based in part on agriculture and commodities, and that this can be compatible not only with forest conservation, but also expansion and increasing per capita incomes for people.

It is an important example, as some commentators and policy makers still maintain that the clearance of the forests is a price we must pay for economic growth, social progress and meeting increasing demand for food. Others argue that economies based in large part on agricultural exports inevitably go hand in hand with rising forest loss. Evidently, neither of these points of view are necessarily true.

Costa Rica, of course, isn't perfect—ecotourism has downsides for the local culture and the emphasis on large-scale fruit exports has led to widespread pesticide pollution—but its distinctive economic strategy has led to very different outcomes than elsewhere in Central America.

Another approach to forest conservation that is leading to positive outcomes across Latin America is via the empowerment of indigenous forest peoples.

10 THE BEST FOREST CUSTODIANS

The most impressive forest outcomes can be gained through the empowerment of indigenous peoples

'The best forest custodians,' Matthew Owen told me, 'are the people who live there.' Owen runs a rainforest charity called Cool Earth and was briefing me about a trip we would take together to visit the Asháninka, an indigenous group in the rainforests of Peru. The Asháninka are the largest of the rainforest tribes in the Peruvian Amazon and one of several communities that Cool Earth works with in its efforts to help the peoples who live in forests to keep them intact.

The Asháninka number around 55,000 people, living in 200 or so remote communities scattered through the Amazon rainforests of Peru and across the border into Brazil. Most of the communities maintain a traditional way of life, entirely dependent on the forest, where they forage, hunt and fish for food, and grow a few garden crops for subsistence and trade. Our destination was a cluster of what Matthew called 'reasonably accessible' villages on the Rio Ené, a tributary of the Amazon. Bounded by the formidable Vilcabamba mountains, it is very remote. To get there, he explained, you travel from Lima across the Andes, then take a truck to the (literal) end of the road, and set off down the river. When you reach Asháninka territory, you are in largely

Marcial Sinchicama, an Asháninka elder, in a traditional *cushma* robe
(Alicia Fox/Cool Earth)

undisturbed rainforest, with jaguars, tapirs, spectacled bears, giant mahogany trees, swarms of colorful butterflies, parrots, hummingbirds and a multitude of other bird species.

'Asháninka' means simply 'The People.' Their home is remote but, like most indigenous societies in South America, they have endured many centuries of disturbance, pressure and threat. Even during pre-Colombian times Inca raiding parties came down from the mountains to take them as slaves, and from the 1530s the conquistadors did likewise. Later there were attempted Christian missions. They all faltered. For a time the missionaries believed they were making progress, but in 1742 the Asháninka rose up and killed all the colonists in the region. After that, the Spanish colonial government, and the independent Peruvian state that followed it, lost control of the eastern slopes of the Andes and the lowland forests beyond.

In the mid-nineteenth century, however, land-hungry settlers encroached more and more into the periphery of the Asháninka's lands, as did the rubber-tapping industry. Some 20,000 square kilometers of Asháninka territory, along with the main rivers, was awarded to a business venture called the Peruvian Corporation, established in London in 1889 as one means for Peru to repay its debts—a prelude of sorts for the so-called Structural Adjustment Programmes implemented a century later by the IMF. The Asháninka people were used as laborers and lost an estimated 80 percent of their population.

Unlike many indigenous groups, the Asháninka survived, often settling deeper into the forest, outside government reach. In the 1970s, when Cool Earth's Dilwyn Jenkins first met them, as a student anthropologist, they had regained an almost entirely traditional way of life in the forest. But then came further incursions. First a wave of colonists, encouraged by Peruvian government policies aimed at opening up the Amazon for landless people, forced the community into more remote areas of the Ené valley. Then the *narcos* arrived, using airstrips cleared from the dense forest to export coca, which grew well in these remote forest areas. And things got worse. In the late 1980s and early 1990s the Asháninka faced new threats from a Maoist terrorist group called Sendero Luminoso ('Shining Path'). This ruthless movement aimed to establish a radical communist regime across Peru, using methods reminiscent of the Khmer Rouge in Cambodia. The Asháninka had the misfortune to find themselves on the front line of the conflict, watching light aircraft arrive with machine guns, artillery and ammunition and the *narcos* who raised the money needed to fund the Senderos' brutal revolution.

Sendero Luminoso seized a huge territory, including Asháninka lands, which they saw as an escape route from counter attack by the Peruvian military. Despite a brutal campaign of intimidation, including taking their children as hostages, again the Asháninka resisted, fighting guerrilla actions. Sendero finally collapsed when, in 1992, its founder and leader Abimael Guzmán was captured, in hiding in Lima. The Asháninka suffered terribly. It is estimated that some 6,000 died, 10,000 were displaced and 5,000 taken captive. Thirty to forty Asháninka communities were entirely destroyed.

The culture of these tenacious people is resilient, though, and today their numbers are rising. They also have some level of protection. In 2003, after years of advocacy, they gained legal rights to some of their ancestral territory, with the establishment of the 1,840-square-kilometer Asháninka Communal Reserve. This covers one of the most pristine and biologically rich ecosystems on Earth, and in theory its reserve status bans new settlements and logging and restricts the expansion of agriculture and livestock. Although the Asháninka people weren't granted property rights, the state recognized their traditional access to the forest for subsistence-based activities.

This arrangement would, it was hoped, protect what remained a near-pristine ecosystem. But, of course, the reality on the ground wasn't so simple. Loggers continued to make inroads around Asháninka territory, offering their communities money for prize trees—in the wake of which came inevitable forest loss, forcing them deeper into the forest. And the Sendero Luminoso had never quite gone away, becoming instead *narco-traficantes*, pressuring the Asháninka to allow them access to transport cocaine. The area, today, is said to be the source of about one fifth of the world's cocaine.

It was in this context that Cool Earth set out to offer support.

Empowerment on the front line

Although Owen had suggested our journey would be to reasonably accessible communities, it was quite an undertaking. First a drive over a high pass over the Andes, where our five-person party struggled for breath, then a winding (and sometimes terrifying) road descent through the agricultural fields of the highlands, with their crops of quinoa, potatoes, barley and wheat that had long since replaced the old temperate rainforest. Finally, an overnight stay at Satipo, before driving on to Puerto Ocopa, a dusty river-port town from where a motorboat took us along the broad channel of the Rio Perené.

A short way downstream we were stopped at an army checkpoint, wary of the *narco-traficantes*. Further on and our boat turned into a tributary—the Rio Ené. From the confluence of the two rivers we headed for about 120 kilometers upstream and deep into the Asháninka lands. Flanked in places by broad flood plains, much of the course of the Ené lay between steep banks that rose to ridges, then to hills and to faraway mountains, all covered with a continuous dark green canopy of dense rainforests

The Rio Ené, Peru. The main route into Asháninka territory (TJ)

running to the distant horizon. To the east lay one of the largest tracts of tropical rainforest left on Earth, stretching across Peru's vast lowlands and far into Brazil. Beyond the distant ridges were tribes that were still living in voluntary isolation.

The river threaded on through rugged Andean foothills. On the banks and flood plains lay the carcasses of great trees thrust downstream on wet season deluges following their fall into the swirling waters on landslides. Flocks of parrots crossed the river and turkey vultures flew intricate maneuvers. Landslides on the steeper slopes and the tumbling nature of the river revealed a young geology. This was the top end of the biggest river system on Earth, with the mountains and hills giving rise to rivers taking Andean sediments first laid down in an ancient seabed. Looking upstream I could see towering white clouds and marveled at how the forest that lay between the river and the mountain ridges was powering the whole system.

Near Satipo we'd seen how a recent road upgrade had already elevated the pressures on the forest. A tarmac surface had been laid about five years previously and alongside it new farms had been established. Many hillsides and valley bottoms had been cleared to make way for bananas, pineapples, manioc (yucca, it is called there), oranges and other crops. A huge truck carrying the cylindrical sections of the trunk of a giant rainforest tree also confirmed the utility of the new road in getting timber out from an otherwise remote and rugged landscape. The frontier of deforestation was heading toward the Asháninka's reserve and if they were going to keep it intact then they'd need to be equipped not only with a legal designation from the Peruvian government, but also the skills and means to hold the line themselves.

That was why Cool Earth was there. As our boat made its way up the Ené, I asked Owen what his group hoped to achieve. He sees Cool Earth essentially as a practical think tank, testing ideas that might be scaled up. 'Indigenous groups are difficult to work with,' he explained. 'There are cultural and language constraints that make it much harder to be effective than it would be to, for example, organize a moratorium on soya production.' He thought such campaigns had been very good at stopping industrial deforestation but less effective in combating smaller-scale degradation, caused by the spread of settler agriculture. This was an increasingly potent threat here on the front line.

Around 7,000 Asháninka people were spread through the landscape that we had passed in the boat, in loose villages of dwellings and gardens surrounded by dense rainforests. Cool

Earth's job, Owen said, was not to persuade the people who lived there to save the forest, but to offer them options. 'The people might have the will to protect the forest, but very often they lack the means. We never try to persuade people to do anything. If you did succeed in convincing them of some particular idea, then you might just create some form of dependency that falls apart as soon as the funder withdraws.' His team's job, he said, was all about empowering communities to have the ability to say 'no' to loggers, *narcos* and settlers.

He explained that Cool Earth also had teams working in rainforests in the Congo and Papua New Guinea, and while each situation was different there were common problems. 'If you are a community dedicated to protecting the forest, the reason you'll lose it, or the big trees growing in it, is because you are very vulnerable and have few other options.' A key strategy, he said, was to make sure that a better income could be derived from what people were already doing, or could do, sustainably. Cool Earth had helped find markets for the cocoa and coffee grown by the Asháninka. 'They were growing these crops already and they are world-class quality. But they sell them for knockdown prices to river traders, and sometimes they struggle to prevent mold, lowering prices further. We've helped with those factors, in ways that don't depend on our support.'

The boat plowed on, its whining engine straining against the push of the river. Sometimes the pilot would slow down to a snail's pace to navigate shallow boulder strewn stretches. Pied lapwings, sandpipers and terns watched from sandbanks as we passed. After many hours the boat arrived at a small riverside settlement serving a community called Cutivereni, or Cuti for short. We scrambled up a steep muddy riverbank to find a few wooden buildings. A group of children and dogs stood at the top and observed our arrival.

This remote spot was once connected to the outside world with an airstrip, but the community had let it grow over, fearing its attraction for the *narcos*, and this had been reinforced by the Peruvian military introducing a no-fly zone. The only way to reach beyond the expanse of forest that surrounded Cuti was on foot, by boat or short-wave radio. The radio provided a link to the Cool Earth team in the event of an emergency.

In addition to settlers looking for places to grow crops, who'd already reached parts of the opposite bank of the Ené, illegal logging remains an ever-present threat, for valuable mahogany trees grow to massive proportions in the Asháninka's forests. The drug producers, too, clear remote areas of forest to grow coca and intimidate and threaten the people.

Deep forest

The day after arriving at Cuti we set off on foot deeper still into the forest toward a community called Taroveni. The route involved wading across swift rocky streams, scrambling up banks, sliding down muddy slopes, and crawling along fallen logs that spanned ravines. I frequently slipped and fell, grabbing at branches smothered in brutal thorns that caused lacerations on my hands and arms. After six hours I was overheated, covered in mud, bruised and bitten by insects. Our Asháninka guides were cool, clean and unflustered by what, for them, was just a short, barefoot stroll.

Taroveni, at 510 meters above sea level, was just high enough for coffee to grow, and to host some of the fauna of the lower Andean foothills, including a magnificent group of bright green military macaws that wheeled in formation over the village. On our arrival, the chief called a meeting for the community of about eighty people, who gathered around as we unpacked presents of knives, threads, fabrics, coloring books and pencils. Young children clung to their mothers, watching us with wide

An Asháninka boy uses a banana leaf for shower protection
(Mark Ellingham/Cool Earth)

eyes. A group of young men looked us up and down and at
our piles of gear, no doubt wondering why we'd come to their
remote home deep in the rainforest. People soon began to relax
and smile, and an impressively large quantity of *masato*, a mildly
alcoholic 'beer' made from manioc and sweet potatoes and
fermented with human saliva, was served in half gourds to wash
down the forequarter of a recently killed wild pig.

The villagers' huts were scattered through a large area of
forest above a small river. They were simple affairs—a wooden
frame covered with a thatch of palm leaves. On the ground

the Asháninka did day-to-day tasks, such as cooking, weaving cotton fabrics, making mats from palms and tool making, and stored their few possessions, including their prized knives and metal pots. Up above was a sleeping platform. Even during the middle of the day these simple structures were surprisingly cool.

Most of the women and children and many of the men wore traditional home-made cotton smocks called *cushmas*—plain colors for the women and stripes for men. On their faces they'd painted streaks and dots of red dye taken from the seeds of achiote shrubs. The people looked healthy and well fed, although as I looked around at the community I saw no old people. Owen told me that Asháninka didn't usually know how old they were. Life expectancy was thought to be around the early-forties and

Traditional Asháninka buildings at Alta Coveja (Cool Earth)

the average age lower because of the high level of infant death, often from diarrhea. Family sizes were quite large, commonly four to six children, in part because the Asháninka still feel a need to rebuild their numbers after the conflict (and also a reflection of the absence of contraception in these far-flung communities). And the communities remain peculiarly vulnerable to disease. One village close to Taroveni had been hit by a flu epidemic and had sealed itself off. More than 500 years after the first arrival of Europeans, the effects of contact were still playing out.

The Asháninka population was nonetheless rising, sustained by those food gardens scattered between the huts and around the village. They were prepared on rotations, using fire, well controlled within the confines, and provided a basic diet of manioc, sweet potatoes, bananas and plantains. The community kept a few chickens, but most of their protein came from the forest and rivers where they hunted for meat and fish.

Gardens and coffee

Cool Earth and others working with the communities before them had concluded that increasing the productivity of smallholder gardens was key to the future of the Asháninka forests. The community already produced crops that could be sold for money, enabling supplies to be bought from outside to augment their forest subsistence. By offering simple but effective advice on how best to increase yields of coffee and cocoa, Cool Earth's local team could help increase incomes, helping the community meet the twin challenges of protecting their forests and reducing poverty.

Some of the gardens had been adapted to basic agroforestry methods, with mixed farming under a canopy of trees. Here, the shade was provided by Inga trees, planted to help retain soil nutrients and to protect the ground from frequent rainstorms. Up above, a flock of about 500 white-collared swifts milled around over the forest. One of the Asháninka explained to me that the birds ('etsoui,' she called them) were 'calling up the rain.' Later on a torrential deluge burst over the village, drenching forest and crops alike.

Saul Quinchori, Cool Earth's agricultural advisor, showed me around the gardens and pointed out the coffee bushes, which he had encouraged the community to plant. He was also helping them to focus on soil health and nutrient capture, using composted kitchen waste, pulp removed from the coffee berries before the beans were dried, and chicken manure. As I looked around I could see a rich layer of decomposing organic material. Moist and cool, the soil was protected from the intense tropical rains and direct sunshine by the layers of plants growing in and over it. Saul told me how the soil microbes and shade helped to produce better-quality coffee and, while it was a little slower than regular plantations, under such conditions 'the trees last for twenty years rather than six.'

Saul also pointed out the advantages of a broader biodiversity—how the variety of birds and insects around the gardens reduced the risk of a serious pest outbreak, as no one species could easily get out of control. I had heard the same thing from organic growers in England, and looking around at the healthy coffee bushes laden with berries in a garden that teemed with life it was evidently working. The wildlife diversity also provided a little extra animal protein in the form of moth pupae collected from the seedpods on the Inga trees. We cooked some with the community. Crunchy and nutritious, they tasted like smoky bacon crisps.

Before the community had Cool Earth's technical advice, assistance and connections, the best they could do to earn a little money was to travel miles through the forest to work for meager wages in coffee plantations belonging to settlers from the mountains. Now they were growing their own produce and were in a position to sell it. There was also a strong emphasis on improved nutrition, with coaching on how to grow and cook different vegetables, how to build ponds for fish rearing and construct proper chicken-keeping cages. 'It might seem an odd way to save a rainforest,' Owen remarked, as we walked past one of the fish ponds, 'but bush meat isn't always available, so these fish can be transformative, especially for child health.' Cool Earth had also invested in safe spring water, piped from the hills. While the rivers are generally unpolluted, they carry sediments and sometimes toxic chemicals used by the *narcos*.

Another key step is education to equip the next generation with a knowledge of Spanish and the basic literacy and numeracy skills needed to deal with outsiders. A local team had been hired, including, where possible, Asháninka people, to assist the communities. It was intended that over time this kind of basic improvement would lift several key indicators, including income, nutrition and longevity. Although not on Cool Earth's agenda, experience from elsewhere suggests that these kinds of improvements will all ultimately impact on family size. It was an approach to development that it was hoped would keep the forest intact by offering the people options for improving their lives, and those of their children, without having to sell trees to loggers, or cede land to *narcos* or settlers.

Camera traps

After a couple of days in Taroveni our group set off again, hiking for six hours through dense primary rainforest to the community of Coveja. In addition to social programs, we saw how Cool Earth had encouraged the Asháninka to become involved in conservation research. At Coveja we were introduced to a remarkable man called Jaime Peña, who was sitting with his wife and daughter in the cool of his palm-roof hut, weaving

Asháninka children wearing traditional *cushmas* and face paint
(Alicia Fox/Cool Earth)

traditional cotton cloth. But Jaime was a very high-tech forest
resident. A couple of solar panels powered a short-wave radio,
printer, video camera and laptop, on which he showed us how
he was documenting the wildlife and mapping the trees around
the village where he lived and had grown up.

The equipment Jaime was using included twelve camera traps
supplied by Cool Earth, which he had scattered in carefully
selected locations through the forest. Onto the computer screen
came amazing images of spectacled bears (one with cubs),
ocelots, jaguars, pumas, tapirs, agoutis, coatis, wild pigs and
forest birds. Jaime made regular long treks through the forest

to change the camera batteries with fresh ones charged from his solar panels and to bring out memory cards with new data.

Jaime lives in a forest that still teems with wildlife. He told us that he believes it is safe, right now, but that demand for more material possessions will increase the pressure. He thinks that if the communities can get organized and make good decisions the forest can be kept, and believes it is important that the communities are aware how precious and unique it is, so he often shows the images he's captured to local children.

We walked on through the forest to a community called Tinkareni. Here cocoa production was an important economic activity and high-quality beans were being produced that could be used to make the very best chocolate, so long as their processing and marketing was up to scratch. Simple but very effective solar driers had been installed to get the best from what they grew. Owen explained how effective preparation and storage of the produce were vital in getting to market and securing good prices, and the potential for making money by doing that better was underlined by another member of our group, top international chef Cláudio Cardoso. Cláudio had a passion for the conservation of the rainforests and was already using Asháninka coffee and cocoa in his high-end London restaurant, Sushisamba. He believed their unique produce had genuine potential.

All these different ways of providing basic assistance are tricky things to do well, not least because of the extreme asymmetry of power that exists between a British-based organization, with its knowledge, connections and budgets, and indigenous and largely illiterate communities. Owen told me that the most basic safeguard was to 'work only with communities that approach us for help.' He explained that once this basic hurdle is crossed Cool Earth goes through a process of informed consent. The communities hold meetings to consider whether and how they wish to work with those offering external help and money; they read out Cool Earth's written proposals in the local language and vote upon each decision. They can walk away from the agreement at any time they wish.

The relationship is a quite complex equation to balance, but on Cool Earth's side there are just three simple conditions that they require. The first is that the communities have proper community representation through an elected decision-making body that agrees how to spend money (which helps to avoid funding just going to the elite families). The second is that the communities produce basic annual accounts. The third is that the rainforest canopy in the areas where the communities live remains intact. It is important, Owen explained, that the money Cool Earth invests is spent by the communities: 'The less we do, the better.'

Rainforest road

Despite the growing pressures on the forests in the Peruvian Amazon, the rate of forest loss in Asháninka areas is far lower than outside. But as the outside world creeps closer, so grows the urgency of enabling the communities to prepare for what lies ahead. From the port on the river at Cuti and running alongside the old airstrip, a rough road extends for 7 kilometers into the forest. The Peruvian government insisted on the route so as to foster future development. Some of the Asháninka are worried about what it might mean.

Adelaida Bustamante, the secretary of Tsimi, a local community organization through which Cool Earth channels funds, told me of her concerns about the new road. Her family had always lived in the forest, she said. It was where they got their food, building materials, water and everything they needed. She feared that such a way of life might be endangered once access was improved, and talked of how the loggers know ways to divide communities and convince people to take money for the giant forest trees. 'The loggers claim that it's not a question of selling wood or trees, but is about getting development for the community.' This sometimes works, she said. Loggers offer to pay for prize timber trees in return for building houses or a community center with left-over wood, and then they say the work was more complex and ask for money. The community has no money, so the loggers take more trees. They come back and do the same and increase the debt to the point where the community can lose its one and only irreplaceable asset—the forest, and for derisory returns.

Adelaida was worried, too, about traders, coming down the new road. 'They have another culture and will bring plastics and trash and will contaminate everything, the rivers and the fish. The community has decided that they don't want the traders in the community and that they should stay on the other side of the river, with the colonists from the Andes.' Another threat is a government plan to connect sewage outfalls upstream. 'The river is our water supply, where we wash and catch fish.'

If experiences from other parts of Peru are anything to go by, the Asháninka have good reason to be worried. Chris Kuahara, Cool Earth's project director in Peru, told me about another rainforest tribe called the Awajún, with whom her team also worked. In their case a series of very serious challenges had accompanied the opening of a new road. 'It was not only environmental pressures,' Chris said. 'Down the new road came tobacco, alcohol, cocaine, prostitution and HIV.'

Indigenous control

Adelaida's commitment to making things work out, for both the people and the forests, was encouraging. From other parts of the Amazon, particularly Colombia and Brazil, where the legal recognition of the rights of indigenous rainforest peoples began a little earlier than in Peru, we can find good reason to be hopeful that what Cool Earth and other organizations are seeking to achieve might actually work.

Martin von Hildebrand is a Colombian anthropologist who has devoted his life to helping indigenous people gain control of their rainforest homelands. He first went to Colombia's tropical rainforests in 1972, to a very remote area where the indigenous people still lived a traditional existence. 'I found to my surprise and horror,' he told me, 'that the indigenous people were still being exploited by rubber merchants, with no rights whatsoever. They were effectively slaves, in permanent debt.' He described a situation that sounded like it was still the nineteenth century. 'Their children were taken by force at the age of six to the missionary school and told that everything in their culture— ancestors, beliefs and daily life—was wrong. They learned a little reading and Spanish but ended up being sent back to their communities totally destitute.'

For the rest of the 1970s von Hildebrand spent his time in the communities, living among the forest Indians. 'I learnt about their culture and way of thinking and gained their trust.' During the 1980s he worked for the government and became close to two successive presidents—Alfonso López Michelsen and Virgilio Barco. 'I told them stories about the Indians and the forests. They wanted to support these communities and the protection of the forest. With their help I managed to get their land rights recognized, as well as their rights in general.' In 1990 he set up an organization called Gaia Amazonas and coordinated with others to help the indigenous groups make the most of their new legally recognized rights, in accordance with their needs, traditions and beliefs.

He pointed to the contrast today compared with the early 1970s. 'Forty-five years later the Indians collectively control 26 million hectares of well conserved rainforest, they run their own schools in their own languages, they have their governments and their rights are recognized in the national constitution. They are able to negotiate with national and regional government as equal partners.' Just to put into perspective what's been achieved, the Colombian rainforest now controlled by the indigenous people is an area larger than Britain.

As I saw in Peru, recognizing the land rights of indigenous peoples living across such huge areas of rainforest can be a very effective conservation strategy. von Hildebrand has seen this from his work in Colombia. 'The indigenous people have developed a world view that aims at safeguarding the forest, not only from a spiritual point of view but because their lives depend on it. It is their home, their identity, their understanding of themselves and the world. They are one with the rainforest. They are the rainforest. In Colombia they offer better protection for the rainforests than the National Parks.'

In other parts of the world, too, researchers have discovered how community-managed forests generally have lower and less variable rates of deforestation than other kinds of protected areas. In Brazil the recognition of indigenous rights has played a vital role in holding the line to slow down or halt deforestation. At the Oslo headquarters of the Rainforest Foundation Norway, then director Dag Hareide showed me a map of the Brazilian Amazon. His organization, and its British and US counterparts, focus on the rights of people as the main means to conserve forests. He pointed out how in the huge swathe of deforestation that had spread across the eastern and southern portions of the Brazilian Amazon during the last three decades, large areas of remaining forest had a high correlation with areas controlled by indigenous people. The Rainforest Foundation's UK office is seeking to replicate this kind of progress in Central Africa, although in that part of the world the situation is rendered more difficult by the absence of any basic legal recognition of indigenous rights.

Even in Brazil and Colombia, however, the situation remains complicated and the recognition of land rights is clearly not a simple, one-stop solution. In both those countries, and as I saw in Peru, the implementation of the law has been slow and complex. von Hildebrand says that one reason for this is in large part intercultural. 'National governments want the indigenous government to be like them, to have a president and secretaries of finance, education, health, environment and all the rest. But the indigenous people have their traditional government based on shamans, communal headmen, ritual leaders, gardeners, hunters and others. The indigenous economy is based on sharing, reciprocity and solidarity, not competition, individualism or the accumulation of financial wealth.' There are also ongoing disputes, as governments exercise rights that they retained when Indian lands were allocated—for example, in exploiting mineral deposits such as oil, which very often coincide with sacred sites. 'The laws protect both sides, so we go to court often.'

There is also ongoing violence, often associated with mining. In September 2017 it was reported that ten members of an uncontacted tribe in the Brazilian Amazon, near the border with Colombia, were massacred by gold miners. The incident took place in the Javari valley—the second largest indigenous reserve in Brazil—and led Survival International to warn that, given the small sizes of uncontacted Amazon tribes, it could mean a whole group would be lost. It was one more worrying consequence of the new government of President Temer, under whom funding for indigenous affairs has been slashed. FUNAI, the main official agency looking after indigenous rights, and which had helped us in our campaign against illegal mahogany logging in Indian reserves, had closed five of its nineteen bases used to monitor and protect isolated tribes, and prevent invasions by loggers and miners. Three of those bases were in the Javari valley, which is believed to be home to more uncontacted tribes than anywhere else on Earth.

As von Hildebrand and others point out in relation to their work supporting indigenous communities, the questions that come to the table go far beyond the conservation of the forests and their wildlife. Also involved are the rights of people. 'It is a fundamental right that all peoples own their land, administer it and consequently control it. The indigenous people are no exception. It is an ethical question as much as the conservation of our planet.'

And it is important to remember that this social dimension of forest conservation is not only about tribal peoples. Other societies who are not descended from the original inhabitants depend on the forest too. Groups that originated from slaves are among those living in and adjacent to forests, and dependent upon them, including parts of Colombia, Central America and Brazil. There are also the people who live around the edge of the remaining forests, and who in some places are the reason why it continues to be cleared to make way for more farmland.

Notwithstanding these social and historical complexities, the recognition of indigenous land rights has been transformative. In tandem with areas protected for wildlife, such as National Parks, a huge contribution to slowing forest loss has been made, and the benefits arising from that don't only accrue to the people who live in the forests of course. As rainforest scholar John Hemming told me, 'Rarely have so many people owed so much to so few. The many are all the rest of the world and the few are the indigenous forest people, who in Brazil alone number only about 200,000, but are looking after 25 percent of the world's remaining rainforests.'

11 TEMPERATE ZONE RAINFOREST

Rainforests also exist at cooler northern and southern latitudes—they are distinct but connected to their tropical namesakes

Before departing the New World rainforests we should pause to note ecosystems in the Western Hemisphere's cooler zones. There are remarkable wet temperate latitude forests in both North and South America. Their original extent was far more limited than the tropical rainforests and much has been logged and cleared. But those patches that remain are among the most spectacular ecosystems on Earth, with trees even bigger than those that soar through the canopies of tropical rainforests. The most impressive and best known are the giant redwood forests of northern California and almost equally tall conifers of the Olympic National Park in Washington State, which are mirrored by coastal rainforest in Chile and Argentina. But temperate rainforests are scattered across the globe in coastal areas with moist oceanic climates, including New Zealand, Australia, Japan, China, Norway and the British Isles, where tiny patches remain, including in Wales.

Pacific Northwest: Hoh

The drama and majesty of natural temperate rainforests can be witnessed along the Pacific Ocean coast of North America from northern California to southern Alaska. Those parts that remain unlogged or clear-felled for timber and paper production are breathtaking for the sheer size of their trees, some of which are hundreds of years old and reach over 75 meters in height and 18 meters in circumference. These huge old conifers include giant redwoods (sequoias), western hemlocks, Douglas firs and Sitka spruce, all growing in abundance due to plentiful water coupled with mild temperatures.

One summer day in 2013 I hiked through an especially fine example in the Olympic National Park known as the Hoh Rainforest. I found huge trees covered with mosses, ferns and lichens, creating a jungle-like impression. Unlike in tropical rainforests, however, decomposition goes slowly, with fallen trunks of the giant trees persisting on the forest floor for centuries. During the long process of recycling, new habitats are created and these provide for whole communities of species, ranging from mosses and lichens to frogs and mammals.

Raised off the ground, and thus a bit closer to the light, tree seeds that germinate on top of these immense fallen logs have a better chance of growing to maturity than those falling into the dark undergrowth beneath. Their roots grow around the rotting trunks and down to the ground. Hundreds of years after seedlings had germinated on a fallen trunk, the trunk finally rots away to reveal straight rows of mature trees standing on stilt-like roots, with a tunnel below where the old log once lay. This feature is called a colonnade, and as I looked through the forest it became apparent how important an aspect it was for the regeneration of the ecosystem. The straight rows of mature trees gave the impression of having been planted by human hand, yet this was the work of nature, using the ruler of a thousand-year-old tree stem to lay out the next generation of giant conifers.

There were deciduous trees, too, and like so many of their tropical rainforest counterparts these are clothed in epiphytes. Some big-leaf maples were so heavily weighted with ferns, mosses and drooping curtains of lichens that huge branches had snapped off. Above the broadleaved trees the straight trunks of the huge conifers soared skyward to converge in a distant canopy, where the sparse needle-clad branches seemed far too modest to collect the solar energy needed to sustain growth.

These forests are home to a number of specially adapted species, many of them threatened with extinction, including, remarkably, little seabirds. Marbled murrelets are—in common with puffins and guillemots—auks. Such birds are creatures of the open sea that come to land only to lay eggs and rear young, generally in colonies on cliffs or inaccessible islands. This little auk is different, however, for it nests in the topmost branches of the tallest of those huge conifers. Sometimes they fly 80 kilometers from the sea to reach these forests to lay their single egg (a fact confirmed in 1974, when a climber found a chick in the top of a forest tree).

As I wandered through the forest wondering why a bird that was otherwise supremely adapted for an ocean-going life had abandoned the cliffs for the treetops, I noticed how open areas created by the crash of a massive trunk were, unlike more shaded areas, rich in shrubs, grasses and herbs. These patches were in turn kept open by browsing elks. Never far away was water, standing in pools and tumbling chilled between moss-smothered boulders, eventually merging into great rushing rivers, through which salmon navigate their way from the Pacific Ocean to spawn in the headwaters.

The majesty of these remaining temperate rainforests was in contrast to recently replanted areas that lay just outside the National Park. There amid the slender stems of trees planted

Sitka spruce. Hoh Rainforest, Olympic National Park, USA (TJ)

about twenty years before, I saw the huge stumps of trees cut during the 1990s. The young trees were ultimately destined for pulp mills and timber yards, as had been the ancient trees that preceded them, which had been clear-felled in the natural forests. Greenpeace and others had mounted campaigns to protect the North American temperate rainforests, although on this occasion had not prevailed. My thoughts drifted back to the 1992 Earth Summit and talks in which the rainforest countries had told the Western nations about what they saw as their right to development and how they intended to exploit their forests for economic growth as the rich countries had done decades or centuries before or, in the case of the primeval forests on the Olympic Peninsula, during the same decade, when the tropical rainforests were so high on the international agenda.

Chile's ancient trees

The remarkable Pacific Northwest ecosystems are mirrored in the Southern Hemisphere, where comparable forests are found along the coast of Chile and into Argentina. As well as being characterized by immense trees, including a number of evergreen broadleaved species that grow alongside deciduous ones and conifers, they have a rich ground layer of ferns and bamboos. They are at their most magnificent in the raw cold wet climate of Patagonia, where the Valdivian rainforests grow in coastal areas, and through the Andean valleys of southern South America, which are closer to Antarctica than to the equator.

Unlike the rainforests of the western US and Canada, the origin of these South American ecosystems is the ancient southern continent of Gondwanaland. That fact is confirmed by how many of the tree and other plant families found in their cool climes are also present in the temperate rainforests of New

Zealand and Australia, including monkey-puzzle tree relatives of those found in the tropical rainforests of New Guinea. Today the Chilean temperate rainforests are isolated from other forests by grasslands, deserts and mountains and host a range of unique species.

Some of the trees in these South American temperate rainforests grow to immense proportions. One is the alerce. It looks like the giant sequoias of California and rivals the bristlecone pine for longevity, with some specimens showing growth rings revealing their age to be in excess of 3,600 years. They are the largest trees in South America, occasionally growing to above 70 meters high, with trunks of 5 meters in diameter. Charles Darwin visited these forests and recorded one specimen with a trunk 12.6 meters across. The biggest trees fell victim to subsequent logging, with the huge specimens seen there in the early nineteenth century taken for structural timber, including for Victorian London's building boom.

Living alongside these giants of the plant world are some miniatures of the animal kingdom. The southern pudú is one of the smallest deer in the world, while the kodkod is South America's smallest wildcat. Today, both of them are among species threatened with extinction due to the loss of native forest to cattle pasture, crop fields, logging and tree plantations.

While it is copious rains that sustain the temperate rainforests of Patagonia, further north are forests that are kept moistened by fog. These include woodlands that grow on the coastal mountains facing the Pacific Ocean in central Chile. This part of the country has a Mediterranean climate and for most of the year it's very dry, yet a few slopes harbor lush temperate forests with a rich ground flora. In October 2017, in the southern spring, I visited such a place on the hills behind the coastal towns of Zapallar and Cachagua.

It was immediately clear that this was a very special ecosystem, full of colorful flowers and songbirds. It is not only very diverse but also unique—a relict of a post-glacial ecosystem that combined elements of both tropical and temperate vegetation. Like other forests its character was determined most fundamentally by how much water arrived there. And although only about 30 centimeters of rain typically falls there each year, another 1.2 meters comes in the cool mists that so often envelop the hills.

I walked through the forest in thick fog and learned from my local guide, naturalist Max Correa, how such conditions were generated by the Humboldt Current, the flow of cold water running north from Antarctica along the western side of South America. It seemed fitting that this ocean current was named after the scientist who first wrote about the connections between water, atmosphere and forests.

Most of the forest's larger trees had been taken centuries before for the manufacture of charcoal, but they were beginning to regenerate, with some help from local conservationists. One reason assistance was needed was because the seeds of the bigger native trees would have been dispersed by now extinct megafauna, such as giant armadillos and ground sloths. One of the conservationists involved in the program, Max's sister-in-law Clarita, walked with us in the forests and explained how she was collecting seeds from threatened species such as the belloto del norte (*Beilschmiedia miersii*), and experimenting with different ways of germinating them in a greenhouse. When the seedlings get going she plants them back into the forest, some of which is now protected.

Although literally on the other side of the world, the forest on that hillside overlooking the Pacific Ocean reminded me of Britain in springtime. Indeed, some of the birds sounded quite similar to those that visit western Britain in summer—one song reminded me of a wood warbler and another a tree pipit. Some of the birds, however, were very different, including the giant hummingbirds that had migrated to nest and rear their young in these temperate latitudes from tropical wintering grounds further north. They'd just arrived and in the spring sunshine, borne on long wings and forked tails, they hunted for insects, their dancing aerobatics a marvel of precision flying, and took nectar from the flowers of pink torch bromeliads.

Connected rainforest worlds

Those beautiful hummingbirds were a reminder that, while temperate and tropical rainforests are geographically remote from one another, they are nonetheless connected. This was highlighted for me when walking in the Olympic National Park, in the mountains above the Hoh valley, where I saw hummingbirds buzzing between flowers against a backdrop of snowy peaks and glaciers. They were rufous hummingbirds, a species that makes a nearly 6,000-kilometer journey twice each year between its wintering range in the woodlands and forests of Mexico to America's Pacific Northwest and even up to Alaska.

In Central and South America it is possible to see many other bird species that migrate long distances between areas of temperate and tropical forest. In the lowland rainforests of Costa Rica I watched a wood thrush, the official bird of Washington, DC, bathing in a tiny stream. In deep shade on the forest floor, the exquisitely spotted bird splashed and fluttered its wings. It reminded me of the song thrushes that frequent my own garden in England, and is indeed a close relative. Unlike the resident songsters that grace our patch, however, this bird was a long-distance migrant between North and Central America.

There are around 250 kinds of birds that make an annual return journey between North America and the New World tropics and, alongside thrushes, hummingbirds, tanagers and flycatchers, are many species of wood warblers. These birds are evolutionarily distinct to the warblers found in Europe but, like those, are mainly migratory and depart the cooler temperate, boreal and Arctic zones to take winter refuge further south, including in the tropical rainforests. While looking for birds in Central and South American forests I've seen quite a few of them—yellow warbler, black-and-white warbler, chestnut-sided warbler, Wilson's warbler and black-throated green warbler among them. Then there are those named with big clues about their long-distance movements, such as Kentucky and Tennessee warblers.

Old World travelers

These American travelers are not alone in making journeys between tropical forests and temperate ones. Each year, a day or so either side of 8 May, birds return from Africa to nest in a cavity by the edge of the roof in our Cambridge house. These long-distance travelers are among the most remarkable of birds, more tuned to an airborne existence than any other. They are swifts and spend part of their year above one of the biggest areas of tropical rainforest wilderness. Then, in summer, they bring a touch of tropical warmth and high-speed exhilaration to the cool and tamer temperate skies of Europe and Asia from Ireland to China, their screaming flocks wheeling through the summer skies, often over large cities.

Our knowledge of the movements of migratory birds like these was, until recently, quite patchy. Today, we have new technology in the form of tiny light sensors that record the time of day and length of daylight and thus enable longitude and latitude to be calculated. When fitted to birds, these devices reveal precise information as to their movements.

Tracking birds with these tools shows that all of the swifts from across their vast summer breeding range, from Shannon to Shanghai, fly in the autumn to Africa, where they congregate over the Congo Basin rainforests. They spread out from there and then disperse into southern Africa, but that is where they all first arrive. One part of the year they are thus feeding on the clouds of flying insects that rise over the tropical rainforest, and another in the damp cool skies of the temperate zone, some going north to breed beyond the Arctic Circle.

Cuckoos are another bird evocative of the short spring and summer in northern Europe. These too have been tracked to the Congo Basin rainforests, with some of them going to an area of forest that is known to be important for a population of western lowland gorillas. The new tracking technology has revealed how English ones get to and from their winter home in the tropical rainforests via Spain, and Scottish ones via Italy. Another species that treks back and forth from the British Isles to the African rainforests is the wood warbler. These delicate little yellowish birds, with staccato song and a preference for moist western woodlands, go back and forth to West Africa, where they penetrate to the edge of the moist Upper Guinea Forests.

It's not only forest birds that are reliant on the rainforests, but also wetland species. Traveling by boat in the mangrove forests of Sumatra in Indonesia, a familiar bird caught my eye, with its characteristic stiff-winged flight and tail-bobbing walk when it landed; it was a common sandpiper. This is a species that I see on fishing and walking trips along the upland rivers and lakes of Britain (although the birds I saw in Indonesia were their Siberian counterparts). Similarly, the spotted sandpipers that I came across wintering in Central America mangroves are seen during the summer by pebbly lake-shores, ponds and stream sides from Newfoundland across North America to Arctic Alaska, including by those streams in the Hoh rainforests of the Olympic National Park. They are another ecological connection between the temperate and tropical rainforests.

A Welsh rainforest

One little temperate rainforest fragment that is little known even in Britain is a woodland called Ty Canol in Pembrokeshire in western Wales. Like pretty much every other British woodland it has been heavily modified by people, and for a very long time. Just how long is underlined by the fact that the area is famed for the quarry from which the builders of Stonehenge took some of the stone.

In contrast to surrounding hills that are thinly clothed with vegetation because of sustained heavy sheep-grazing, the patch of woodland that survives here conveys at least some sense of what the land was like before nearly all of it was cleared for different kinds of farming. Among rocky areas and exposed low crags grow gnarled old sessile oak and ash trees. Smothered with ferns and mosses, this is one of the richest sites in the British Isles for different kinds of lichens. About 400 species have been recorded here, many of them very rare.

Lichens were heavily impacted by the pollution of the Industrial Revolution and the profusion of these curious organisms, comprised of a partnership between fungi and algae, confirms not only a steady supply of pure water (it rains most days, and sometimes for days on end) but also very clean air. The moisture that bathes this little temperate forest fragment is delivered from the warm waters of the Gulf Stream, the ocean current that was one more consequence of the tectonic upheavals that joined North and South America, diverting ocean circulation as the rise of the Isthmus of Panama split the Atlantic and Pacific Oceans.

Ty Canol Wood, Pembrokeshire, Wales (TJ)

Walking there in early summer, when migrant birds including wood warblers and redstarts are fresh in from their African wintering grounds and singing on their newly established territories, it is oddly reminiscent of a tropical cloud forest. These globe-flying forest birds seem to choose forest habitats when they go north, inherited from tropical ancestors who established the annual routes several million years ago. So it is that in a Welsh wood in summer the impression of a tropical rainforest is closer to the truth than one might imagine. The story of the tropical rainforests now turns to the places where these birds spend most of their time—the forests of Africa.

PART THREE

AFRICA

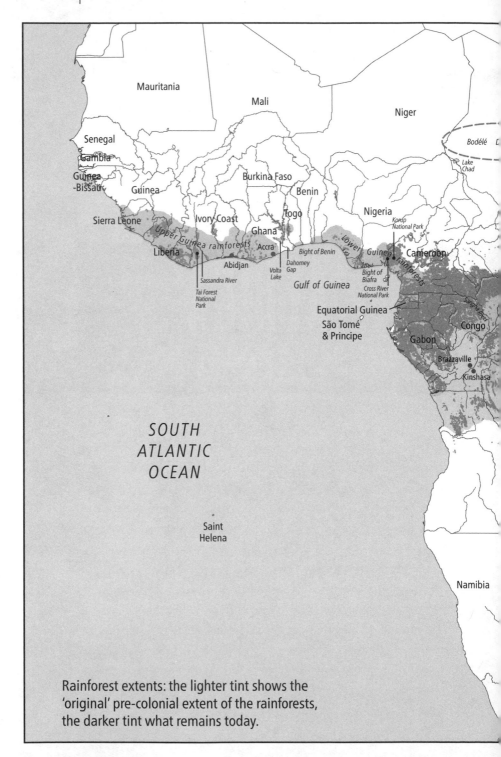

Mauritania

Mali

Niger

Senegal

Bodélé

Gambia

Lake
Chad

Guinea
-Bissau

Burkina Faso

Guinea

Benin

Nigeria

Sierra Leone

Ivory Coast

Korup
National Park

Upper Guinea rainforests

Ghana

Togo

Liberia

Accra

Lower Guinea rainforests

Cameroon

Abidjan

Bight of Benin

Dahomey
Gap

Sassandra River

Volta
Lake

Bight of
Biafra

Congo

Tai Forest
National
Park

Gulf of Guinea

Cross River
National Park

Sangha River

Equatorial Guinea

São Tomé
& Principe

Gabon

Brazzaville

Kinshasa

*SOUTH
ATLANTIC
OCEAN*

Saint
Helena

Namibia

Rainforest extents: the lighter tint shows the
'original' pre-colonial extent of the rainforests,
the darker tint what remains today.

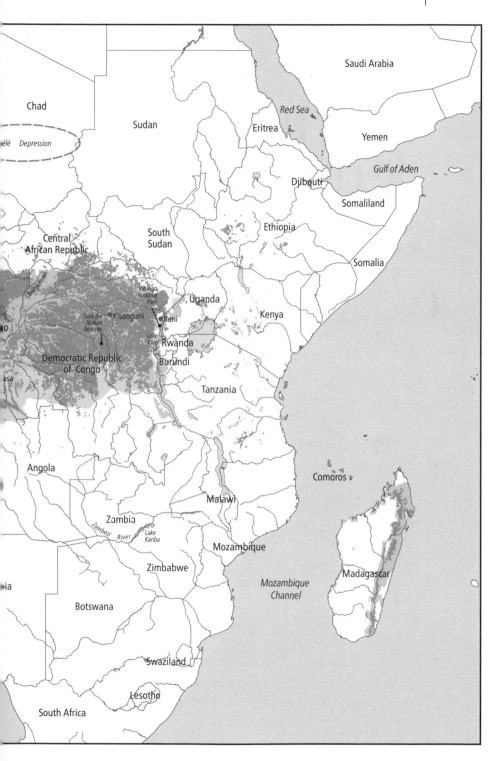

12 LAST FRONTIERS: THE CONGO BASIN

The Congo Basin is the largest rainforest after the Amazon; it is still largely intact but pressures are mounting

The African rainforests—as across the Atlantic—hosted human civilizations long before the first European explorers arrived. Indeed, it is believed people have lived in the Congo basin for about 50,000 years. One of the most famous of all prehistoric human artifacts is the Ishango bone, found by Belgian explorers in 1960 in what was then Belgian Congo (now the Democratic Republic of Congo, or DRC). This baboon femur, around 20,000 years old, bears carvings believed to be some kind of counting system, the earliest example of humans inscribing mathematical calculations—at a time long before there were any humans at all in the Americas.

Since those prehistoric times, the moist forests expanded and contracted in response to climatic change and the wetter and drier periods that came with such oscillations. Between about 4,000 and 2,000 years ago a drier time led to a contraction of the dense humid forests. During this period, groups of stone-using hunter-gatherers gave way to colonization by Neolithic farmers moving south from the Sahel region. They practiced rudimentary slash-and-burn methods, and because of the drier conditions were able to penetrate further into the Central African forests. Around 2,800 years ago iron tools spread among the farmers and that likely had a significant impact on the forests, not only because of the greater efficiency with which land could be cleared of trees, but also because of the demand for charcoal needed to smelt iron ore. This Iron Age civilization is believed to have reached a peak some 1,900 years ago before suffering a population crash around 400 CE, possibly due to outbreaks of Ebola disease. As the human population contracted, so the forest expanded, until a wave of metal-working colonists arrived around the year 1000. Human numbers gradually increased over the next 500 years, when the population crashed once again. This time, we know the cause—the Atlantic slave trade.

Slavery and Africa's forest people

Between 1418 and the 1470s Portuguese ships made a number of exploratory expeditions to map the oceans that lay to the south of their home ports, including the coast of West Africa. As they came to know the new continent they found that fortunes could be made from ivory, gold . . . and people. Through collaboration with local slavers the Portuguese, and then the Dutch, English, French and Scandinavians, found people-trafficking to be a hugely lucrative business. In 1526 a Portuguese ship made the first transatlantic slave voyage from Africa to the Americas, supplying labor for the new Caribbean plantations. By the 1600s the capture and shipping of slaves had become a major business, driven by demand for the huge number of workers needed to keep up with fast-rising consumption of sugar and other tropical produce in Europe. The decimation by disease of the indigenous populations of the Americas further increased demands for labor to be supplied from Africa.

Slavers rounded up people living by the coasts and rivers and transported them in appalling conditions. This was for the most part done by West African natives, although as time went on European buyers staged more of their own raids, so as to reduce costs. Many of those pressed into slavery by this brutal trade died on ships, and many more before they even set sail: perishing on the forced march from their homes or while awaiting sale in so-called slave forts on the coast. In the absence of comprehensive records it is impossible to state with accuracy the overall scale of this barbaric business but the impacts on the varied cultures and complex political structures of Central and West Africa were devastating. It is estimated that 9–11 million people were shipped out of Africa. It might be that as many again died en route, making a total in excess of 20 million people.

Those that did get aboard ships became part of the slave triangle, whereby vessels heading for the Americas with slaves would be unloaded of their human cargoes before setting sail to Europe laden with coffee, sugar, tobacco and rice. From there ships would then once more head to the western coasts of Africa, loaded with goods that would be traded for a fresh load of slaves, before setting off again to the West.

The farmed areas among the forests most accessible to the slavers, and thus most vulnerable to attack by raiding parties, became depopulated because people were either captured or fled. This meant that, from the gardens tended by the indigenous people, the forests began to grow back. As with the Amazon, it is possible to use data from satellites to identify regions of forest—for example, those in the semi-deciduous region along the Sangha river that were abandoned. Amazingly, the fingerprint of those social upheavals is also discernible in polar ice. The data gathered from those frosty samples reveal that during the early 1600s there was a sharp drop in the global atmospheric carbon dioxide concentration. Over the course of a few decades it went down by about eight parts per million, and it stayed at that lower level until the mid-nineteenth century, when the large-scale burning of coal got fully underway. The *drop* in the carbon dioxide level was caused by the terrible human toll of slavery in Africa and by sudden large-scale population crashes among the native peoples in the Americas, meaning that large areas of previously cultivated agricultural land quickly returned to tropical rainforest. When that happened, carbon dioxide was extracted from the atmosphere, and on a massive scale.

Congo genocide

The human costs of European incursions into Africa's rainforests continued during the colonial period that followed, most infamously in what became known as the Congo Free State, the personal fiefdom of King Leopold II of Belgium. Today known as the Democratic Republic of the Congo (DRC), this vast area of Central Africa, with the continent's greatest rainforest at its heart, was subject to merciless exploitation during the latter part of the late-nineteenth and early-twentieth centuries.

At first, Leopold's plunder was the collection of ivory, but during the 1890s the focus shifted toward the production of rubber. This was derived from a different species to the plants fueling the parallel boom in South America and came from a type of vine native to the dense jungles of the Congo basin. Mercenaries were hired to drive local people into forced labor. Villages were set production quotas, with various tactics employed to speed up delivery—including torture, arson, rape, hostage taking and murder. First-hand accounts of the brutality were relayed to the outside world, including to the USA by the African American Presbyterian missionary William Henry Sheppard. He wrote in an article published in the *New York Times* in January 1900 about 'Fourteen villages destroyed by fire and plundered, 47 bodies lying around, three with flesh removed and eaten, 81 hands removed to be handed to Belgian authorities.' This was the setting, too, for Joseph Conrad's 1899 novel *Heart of Darkness*.

An international scandal did eventually ensue, due in part to Sheppard, and political pressure forced the Belgian government in 1908 to wrest control from King Leopold and to establish a civil administration. But the cost had been enormous, with an estimated 10 million native people murdered or worked to death by Leopold's men.

Congo Basin rainforest

Covering an area of more than 2 million square kilometers, the Congo Basin rainforest is exceeded in scale only by that of the Amazon. And like the Amazon there is a mosaic of forest types shaped by the influence of rivers, geology, climate and soil, including a range of swamp and flooded forests and those growing on drier ground and uplands. The rainforest here extends from the Lower Guinea forests of Nigeria and Cameroon, with which they are contiguous, south to Gabon and Congo and across the Central African Republic to the mountains that flank the Great Rift Valley in Uganda, Rwanda and Burundi. However, it is the Democratic Republic of the Congo that has by far the biggest share.

Following the depredations of the rubber boom, the rainforests of the Congo Basin continued to supply resources to world markets, including timber, diamonds, metal ores and agricultural crops. But until recently the pressure on the forests was modest and the rainforest relatively undisturbed. However, with rising global demand for the Congo's particular resources, over the last couple of decades there has been rapid deforestation and exploitation. This has been exacerbated by local pressures arising from the growing human population and made more feasible by the ebbing of DRC's long civil war and conflict.

Most of the tens of millions of people living in the Congo Basin depend heavily on the rainforests for gathering wild

Aerial view, Tropical Rainforest, Democratic Republic of Congo (Getty)

products that they need for food, raw materials and wood fuel, as well as land upon which to grow crops. It is the more accessible drier forests and forest edges that have been most vulnerable to being degraded and cleared, as well as those closest to the navigable rivers, including the Congo and Ubangi, and of course those close to roads. But as the population continues to rise, so do the pressures on the forests. These pressures have in some areas recently increased, such as around Lake Kivu in the eastern Congo Basin, where in a post-conflict situation people are moving back to villages from refugee camps, increasing the demand for wood and meat from the surrounding forests.

It is not only the rural populations that make demands on the forests but also the growing number of people living in the substantial cities. The vast metropolis of Kinshasa faces its smaller (but still large) neighbor Brazzaville across the broad lower reaches of the Congo river about 150 kilometers from where it meets the Atlantic Ocean. Kinshasa is the capital of the DRC, and Brazzaville that of the Republic of the Congo. The millions of inhabitants of these huge cities drive demand for a variety of forest products, including culturally popular 'bushmeat,' which comes from wild animals, including endangered species such as gorillas.

Assaults on the last frontiers

Like the region's indigenous people, the Congo Basin's gorillas are under pressure from the disturbance and fragmentation of the forests caused by logging activities. Logging, although often called 'selective,' can be likened to the bombing of a city. Some functions of the forests still remain, while others are seriously disrupted, taking time to recover to their original state. If the forest gets repeatedly logged, then the disruption continues, perhaps to the point where some functions or characteristics of the forest don't get a chance to recover or replenish.

Most of the Congo Basin rainforests have already been allocated for logging activities and on top of state-sanctioned timber extraction illegal logging is also widespread. In both cases, not only does the disturbance to the forest cause impacts on the wildlife, but the trails cut into the forests to reach the most valuable trees are often used by people to reach previously inaccessible areas, where they clear the trees completely to make way for crops, while at the same time increasing the poaching pressures on populations of wild animals.

Although since 1990 the deforestation rate in Central Africa has been the lowest of any major forest region in the world, the rate of forest loss has recently gone up. One estimate suggests that compared with the period between 1990 and 2000, the rate of deforestation between 2000 and 2005 was about double. The World Resources Institute (WRI) estimated that from 2000 to 2014 the DRC alone lost more than half a million hectares (nearly 5,000 square kiolometers) of forest each year, including encroachment into protected areas such as Sankuru Nature Reserve, the world's largest continuous protected habitat for great apes. Over those fifteen years the total lost was thus about 75,000 square kilometers—about the size of Ireland.

WRI's analysis of satellite data also revealed the progressive fragmentation of the forests, especially along the lines of major roads, and major towns, including Beni in the eastern province of North Kivu, where civil unrest and conflict caused a wave of inward migration. Other new hotspots of deforestation have begun to appear around other major population centers, such as Kisangani, as forests are opened for new agricultural land and for charcoal production to supply the rapidly expanding population with food and energy. Across the tropics road-building has already opened up many previously remote areas of forests, and the number and reach of new roads is set to continue to grow. Researchers estimate that by 2050 a further 25 million kilometers of new roads will be built globally. That would be a 60 percent increase over 2010's total road length and nine tenths of those new roads are expected to be in developing countries, including those with tropical rainforests.

There is also the damage caused by mining activities. The Congo Basin is rich in mineral deposits and beneath the tropical rainforests there are massive quantities of gold, copper, diamonds, cobalt and rare minerals, including coltan and cassiterite, which are in rapidly rising demand because they are needed in the manufacture of laptops, smartphones and other electronic devices. There is also a substantial quantity of oil.

As well as major mining operations under the control of major multinationals, there is a huge informal mining sector that sustains the livelihoods of millions of people. Between them these activities now comprise the largest single economic sector. Although often a cause of forest loss, as trees are cleared to get at the minerals, there is reason to believe that well-organized mining and drilling operations can help protect forests. In Gabon, for example, the high economic value of oil extraction causes a relatively small footprint in terms of forest loss, and has been used to provide financial resources to prevent illegal activities, including hunting. However, in the DRC oil and gas exploration seems unfettered, even being proposed in the montane rainforests of the Virunga National Park, which is of vital importance for the survival of, among many other creatures, the critically endangered mountain gorilla.

Exploration for oil and gas was merely the latest threat to Africa's oldest National Park. From refugees fleeing the 1994 Rwandan genocide to the local charcoal mafia killing gorillas so as to lessen the importance of the Park (and thus the effort being put into protecting it), it has been hard to hold the line here. However, an international campaign by WWF did prevent the British oil company Soco from getting access to the 7,800-square-kilometer UNESCO World Heritage Site. For now it remains a safe haven for its diverse wildlife and continues to enjoy the distinction of being the only protected area on

Mountain gorilla silverback (*Gorilla beringei beringei*), Volcanoes National Park, Rwanda (TM)

Earth where three kinds of great apes live (the other two being eastern lowland gorillas and chimpanzees).

Mining is a major threat to the world's tropical forests—and most of all to those in Africa. Mining projects have often led to international controversy, particularly on the biologically unique island of Madagascar. A belt of moist tropical rainforests along the eastern side of Madagascar was once packed with unique wildlife. That has now mostly gone. In the south there are also areas of unique spiny forest, again packed full of species found nowhere else on Earth. Those too have been depleted to the point of disappearance, including through a huge mining project by global mining giant Rio Tinto.

It was during an investigation into this project in 1994 that my mentor Andrew Lees, Friends of the Earth's campaigns director, met his death. He went to film in the threatened forest and was never seen alive again. Despite an international campaign, the mine was subsequently built, earning Madagascar revenues from exports, but in the process it caused the extinction of countless animals and plants.

Indigenous survivors

Despite the depredations of slavery, the rubber boom and subsequent incursions for timber and minerals, some tribal groups did survive and continue to live more or less traditional lives in the rainforests. In the Congo Basin the Baka are the best known people still pretty much maintaining a hunter-gatherer lifestyle, dependent upon and intertwined with the forests. Other groups include the BaMbuti, Efe and Aka. Underlining the richness of their cultural heritage is the fact that there are still around 210 languages spoken in the DRC alone.

Living at low population density today, however, there are perhaps as few as 170,000 people who remain in traditional forest societies, scattered across the Congo region. Most live in

Baka woman hunting in southern Central African Republic in 2014 (Max Chiswick/WikiCommons)

bands of up to seventy people and tend to be nomadic, moving to new areas of rainforests several times a year. This, coupled with their small numbers, means that their overall demands on the forests are sustainable, as they hunt and gather what they need. Pressures on the forests have increased, however, due to some groups trading with villages, exchanging game and honey in return for agricultural produce, thereby causing the intensity of hunting to rise.

It's not only the game that is under pressure in the Central African rainforests, but also the indigenous peoples that rely upon it. As already noted, logging operations cause not just direct disturbance but access tracks into previously inaccessible areas. This, coupled with the expansion of villages because of rising population, means that more and more people are hunting in the forests, depleting the indigenous people's food supply. Logging camps also bring diseases deep into the forests. Mining, even when done well, can similarly lead to forest disturbance and cultural disruption. On top of all this are official government policies geared toward ending forest peoples' traditional lifestyles, compounded by the fact that no Central African country has recognized any legal land rights among such groups.

It is important to note that, while the forest peoples living in traditional groups are described as 'indigenous,' they are at the same time ruled by elites in national governments who, unlike their counterparts who run most Latin American countries, are descendants of people who lived there before colonization. In most Central and South American countries the populations are descended from either European colonists or intermarriage between them and indigenous people, which creates a different context for 'indigenous' societies to be identified.

In any event, and often irrespective of the ancestral descent of decision makers today, the national governments of the rainforest countries are sometimes wary of recognizing indigenous rights and land title, knowing that this might pose constraints on future development choices. For example, if land title was granted to an indigenous group and then later on a major body of ore was found beneath their land, the forest people might be in a position to block mining operations. The potential for that to happen leads some governments to regard indigenous rights as posing an economic risk.

During the course of lobbying the political negotiations leading to the 1992 Earth Summit in Rio de Janeiro, I was surprised by how strongly African rainforest country governments were set against wording that related to the rights of indigenous forest peoples. A weak statement on 'Forest Principles' that was one of the outputs of the 1992 Summit underlined the tensions at hand. In that agreement, the result of particularly acrimonious talks, there were a few vague references to indigenous people, but set against that was far stronger wording about nations' right to development, and their rights to sovereignty over their natural resources and ability to exploit them.

The 'right to development' reflected a basic political choice about the control of natural resources, and how that control would be harnessed for economic growth and development, in relation to timber, minerals, river flow for power and soils to grow crops for export. The mounting pressures on the Central African forests are one manifestation of how countries have tended to pursue their development goals. While no one can predict the future, by looking at what has happened in Africa's other major block of rainforests we can gain some impression of one possible future fate of the vast Central African rainforest ecosystem.

13 THE LOST FORESTS OF WEST AFRICA

The 'Upper Guinea' rainforests of West Africa are home to many unique species, but their extent today is fragmentary

Rainforests once stretched in an unbroken, 1,500-kilometer belt across West Africa, along the coast from Guinea to Ghana and for hundreds of kilometers inland. These moist forests gave way on their northern fringes to the savanna woodlands of the Sahel, which eventually graded into the Sahara Desert. This region is still known as the Upper Guinea Rainforests and is separated from the Lower Guinea and Congo Basin forests by a stretch of savanna that reaches south to the coast in the little West African countries of Benin and Togo, forming a boundary between the two main African rainforest areas known as the Dahomey Gap.

Although the tropical rainforests in this part of Africa are geographically the closest to Europe, there wasn't initially a high level of forest loss. Some of the goods in greatest demand during colonial times, including gold, ivory and kola nuts, were supplied with a low level of forest loss and, as in Central Africa, the capture of slaves led to forests regenerating. Later, trade in gum copal produced from plantations of *Daniella* trees began to encroach on forest land. Then during the early nineteenth century the colonial powers became aware of the huge economic potential of palm oil.

Cocoa pods at various stages of ripening (Medicaster/WikiCommons)

Palm oil and cocoa

Today oil palms are grown in vast plantations across the wet tropical regions, especially Southeast Asia, and produce huge quantities of oil, used in countless consumer products. In West Africa, however, palms were to begin with harvested from wild plants. Farmers who cleared areas of forest left the palms, which are native to this part of the world, standing in their fields. Given space, they spread easily, creating semi-wild groves that were tended for the oil-rich fruits. In 1820 the Gold Coast (as Ghana was then called) first exported palm oil to Europe. The British encouraged the trade and it became the principal cargo for former slave traders and ships after the abolition of the slave trade in 1808. In the 1850s plantations were established in the Gold Coast, and by the 1880s palm oil accounted for 75 percent of the country's export revenues.

The rise of West African palm oil was accompanied by the establishment of an even more lucrative crop: cocoa. During the latter quarter of the nineteenth century large areas of coastal Gold Coast and Nigeria were cleared to make way for it, and today the moist forest zone of West Africa is the most important region in the world for cocoa production.

The effects on the rainforests from subsequent cocoa expansion can be seen across the region and especially in Ivory Coast, today the world's largest cocoa producer. That country has been losing its forests faster than any other African nation, including from what were on paper at least protected areas. At the time of independence from France in 1960, there were about 140,000 square kilometers of natural tropical forest cover in the country. Today only 10,000 square kilometers remain, the vast majority of it in National Parks and other protected areas, including the Tai Forest National Park (more on which later). The conservation group Mighty Earth estimated that, between

2000 and 2014, Ivory Coast lost nearly 5,000 square kilometers from areas *set aside for conservation.*

In Ghana cocoa has been a similarly powerful cause of deforestation, with nearly 850 square kilometers lost from protected forests during the same period. Some of this deforestation has occurred in the largest remaining areas of chimpanzee habitat in those countries. The West African rainforests have also been heavily impacted by the timber industry.

Logging and war

In 1990 I visited the dockside at Abidjan, Ivory Coast's main commercial city and port, where I spoke to the workers loading logs for export to Europe. They told me that the country had no means to add value to the wood, for example by processing the logs into planks, furniture or window frames. This was why so little economic value was attached to the forests as a sustainable resource and why it was being cut, cleared and replaced with crops. The size of the logs on the quayside said a lot about how far the depletion of the forest had gone, with trees as small as 30 centimeters in diameter piled up on the quays. The bigger and more valuable specimens were long gone.

Not only has the timber industry been a weak contributor to economic development in the region, but it has also been a factor in conflict. From the early 1990s to early 2000s timber revenues helped to fuel the conflict that raged in Sierra Leone. Charles Taylor, president of neighboring Liberia, promoted unrest in order to gain control of Sierra Leone's diamond mines. The long civil war that resulted was marked by appalling atrocities, including slavery, amputations and executions. As the world became increasingly aware of what was going on, there was eventually a ban on the trafficking of so-called 'blood diamonds.'

While that didn't completely stop the trade, it did significantly curtail it and it led Taylor to seek cash from other sources, including timber. Some of the money raised from the expansion of highly destructive logging operations was used to pay for the weapons that perpetuated the conflict in Sierra Leone and funded militias in Liberia. Mature logs appeared again on the Abidjan docks, having been illegally shipped out of Liberia. Internal conflict also caused major damage to the rainforests in Ivory Coast, where during the political crisis of 2010–11 two large protected areas were cleared. Illegal logging, conversion of forests to farming and mining were the causes, while intensive poaching added to the pressures on surviving wildlife.

The Lower Guinea Forests of Nigeria and Cameroon

The situation is little better in the Lower Guinea Forests that run from Benin through coastal Nigeria and Cameroon and south to Equatorial Guinea. These are important wildlife areas. The very wet montane forests in the highlands of Cameroon also possess an amazing wealth of species found nowhere else on Earth, as do the forests that extend to the Gulf of Guinea islands of São Tomé and Principe.

However, all these forests have been highly fragmented and much of what's left remains only inside protected areas. The best known of these are the Cross River Park in Nigeria, where there is a globally important population of chimpanzees and western lowland gorillas, and the Korup National Park, across the border in Cameroon, which is home to one of the most significant populations of rainforest elephants in Africa.

Alongside demand for farmland in these countries, rapid urbanization has taken a toll, including the rise of one of the world's fastest-growing and largest cities—the Nigerian capital, Lagos. And, as in the Congo Basin, the rainforests have not only been under pressure from logging, encroachment and

Fragments of natural forest survive in areas of unproductive agriculture and scrub, western Ivory Coast (TJ)

urbanization, but also heavy poaching, meaning that the tropical rainforests have not only shrunk, but their health and ecological viability has been affected by the decline of wildlife.

Poached forests

Walking through a patch of scrubby farmland adjacent to patches of degraded rainforests in western Ivory Coast, I spotted pied crows, shrikes and weaver birds. A colony of weaver birds had built their hanging grass nests on an intersection of power cables; the black and orange males were hanging upside down, their wings flapping, with scratchy songs advertising half-built nests that they hoped would present sufficient promise to attract a female.

Weavers are quite small birds, about the size of sparrows. There were plenty of them, evidently taking advantage of the grasses and weeds growing in the fields and upon the seeds of which they fed. Those fields were also buzzing with the insects that they fed to their chicks, suggesting a low-level use of agricultural pesticides. Dragonflies sped around the edge of clear ponds that had water with purity born from the scarcity of crop fertilizers. Although it was a largely deforested landscape, there did appear to be quite a lot of wildlife.

But as I traveled through mosaics of farms and forest fragments I noticed no large animals—no deer or antelope, no monkeys, no big birds, not even any fruit bats. The reason, I discovered, was that the people had eaten them. Short of protein and animal fat, the rural communities had snared, shot and trapped pretty much every meal-sized creature they could get their hands on. I could see why. There were few shops selling meat and, even if a good-quality refrigerated supply chain did exist, most people wouldn't be able to afford what it sold. For poor people the logical choice was to invest in a gun, snare or a net and to get hold of what they could from the countryside.

Meat from wild species is a critical protein source for many rural people in the tropics and, as the rural population has increased, so the pressures on animals, including forest pigs, antelopes and primates, has become ever more intense. In some rainforest regions of Africa the threat to many species of mammal arising from hunting now parallels or exceeds that of habitat loss. Countries have enacted laws to protect wildlife and ban hunting of certain species, but in remote rural areas and with little enforcement capacity the effects of these have been limited, especially when some animals are trapped not only for subsistence food but for lucrative urban markets.

For example, the Congo Basin is the only part of the world where all four subspecies of gorilla live. All are at dire risk and the gravest threat to their survival is poaching to supply bushmeat for sale in cities, where among some wealthy elites the consumption of ape flesh is considered prestigious. Many other large animals accessible to poachers are similarly targeted. The more the forests have been opened by tracks driven in to extract logs or establish mines, the worse such pressures have become. For poor people on tiny incomes, additional money from poaching can be very attractive.

Other tropical rainforest animals are caught and shot to fuel trade in skins, medical products and live animals for pets and collections. Between them these pressures are causing the rapid decline of once widespread and abundant species.

The ecological consequences of the removal of mammals and birds from forest systems can cause cascades of effects that can

Women and girls collect firewood a few kilometers from the Tai Forest National Park, Ivory Coast (TJ)

be as profound as they are unpredictable. Indeed, there is good reason to see the conservation of declining rainforest animals as a job that is equally important as reducing and reversing deforestation itself. But behind all the effects of rising demand for natural resources, logging, the clearance of land for farming and poaching have been two even more fundamental factors—rising population and corruption—both of which, of course, feed on poverty.

Ivory Coast poverty traps

Returning to Ivory Coast in 2016, I traveled through rural landscapes where rainforests had recently become farmland. For the most part the farms were subsistence enterprises, run by some of the billion or so people in the world living below the threshold that the UN regard as absolute poverty. With limited access to clean water, power and sufficient food, they struggle to feed and clothe themselves, never mind educate their children.

One possible route out of the rural poverty trap was to move to the city, to live in a shanty town and hope that one day it might be possible to get some decent work. Millions had taken that gamble, and that was partly why Abidjan had about twice as many people as the last time I'd gone there a quarter of a century before. Aside from moving to the city, I was told that many had made the hazardous trip across the Sahara Desert to Libya, and then the deadly journey across the Mediterranean to Europe.

I could understand why they might wish to move. Farming seemed an almost impossible route to a good living. Widely spaced rows of maize were planted in fields where weeds and grasses grew nearly as tall as the spindly corn plants. There were overgrown banana fields and poorly tended plantings of cassavas and yams, suggesting not only a scarcity of fertilizer but also a lack of expertise among local farmers.

A truck laden with a harvest of oil palm heads contrasted with what I'd seen on numerous visits to Malaysia and Indonesia, where in some mature plantations a hectare of palms can produce upwards of 8 tonnes of oil a year. Whereas in Asia the big fruiting heads were thickly studded with swollen amber capsules, here they were smaller and the gleaming oil pods fewer.

There were also some thin pastures where cattle and goats grazed between what were once canopy trees. Standing high on buttressed stems, the forest giants had survived the deforestation and alongside some stumps that remained in the fields revealed how not so long ago the rainforest was still here. Much of the land had become degraded, with its soil badly eroded, covered with scrub, and even where cultivated was producing little food, underlining that the economic benefits arising from deforestation had been very limited.

As might be expected, the pressures on Ivory Coast's forests had become more acute in proportion to the rising number of people. In 1960 the population of Ivory Coast was around 3.4 million people; by 2013 that had risen to 20 million. Even though population growth in the big cities had been dramatic, most people still lived in the countryside, where the majority were dependent on farming. An expanding rural population had unsurprisingly translated into demand for more land. But, while conditions in the rural areas had led to high fertility, they didn't promote long life. Poor housing, a lack of sanitation, disease, intermittent electrical power, limited access to clean water and grinding hard work were among factors that added up to a striking absence of many older people.

The land clearance that had taken place in Ivory Coast was primarily driven by small-scale farmers converting land to meet their immediate needs, or to produce modest quantities of produce from their small plots for export—rubber, but especially

Holding the line where extreme poverty meets a National Park, Tai Forest, Ivory Coast (TJ)

cocoa. The expansion of the cocoa plantations that had sustained so many rural families was not only an historic driver of forest loss but also a current one. Ivory Coast and Ghana between them supply more than half of the world's cocoa and, as well as being the main (or only) source of income for the farmers who grow it and a major source of taxes and foreign exchange for the countries that export it, cocoa is also the basis of the profits of the global chocolate companies whose dividends are paid out each year to pension funds and other investors. All this comes at the expense of the rainforests. As production continues to expand, even the last protected forests are at risk.

On top of land to grow crops, the growing population also look to the remaining forest to supply its fuel. Some of the better-off families living in the villages were able to afford refillable canisters of natural gas for cooking, but many women and girls carried great piles of wood on their heads. Collected from woodlands and remaining patches of forest, this was an additional pressure that not only ate away at the forests but also prevented its regeneration.

Tai Forest: parks under siege

Visiting a cocoa plantation in Ivory Coast I noticed, amid a patch of flowering herbaceous plants, a concrete triangulation post. It marked the edge of the Tai Forest National Park, and signaled where agriculture met the largest remaining block of natural tropical rainforest left in West Africa.

Next to that forest were some of the poorest communities I'd ever seen. The people lived in huts made from mud bricks and roofed with black plastic sheets weighed down with whatever was to hand—polythene bags full of soil, bits of scrap metal. The huts seemed insufficient shelter to protect against the heavy tropical rains, and where the mud had washed out there were holes in the walls. A few homes had corrugated metal roofs, but were not much better. Outside many of these little houses were cocoa beans, drying on grass mats under the hot sun. Without a market for that crop, the people would starve.

Two days before my visit to the Tai Forest I'd met Diarrassouba Adoulaye, director for the Park's eastern sectors. Adoulaye showed me a map of the boundary and the many hundreds of settlements alongside each one, he told me, full of people living in dire poverty. The Park itself had been established in 1926 and, including its buffer zone, covered around 5,000 square kilometers. It was now the biggest and best preserved swathe of natural tropical rainforest in West Africa.

The Tai Forest is one of Earth's richest biological treasures, home to a famous population of tool-using chimpanzees as well as forest elephants and the white-necked picathartes—a strange bird with a bare white head that is found only in the remnants of the Upper Guinea rainforests. There are pygmy hippopotamuses, leopards and zebra duikers (a threatened species of forest antelope) and several species of monkey, including the beautiful black-and-white colobus.

I walked between the cocoa plantation and the park boundary with Lieutenant Assie Roy of the Ivoirian forest department. Smartly uniformed, he was dedicated to the challenging task of ensuring the integrity of the Tai Forest. He explained how his small team of men had planted rows of rubber trees along this section of the boundary to show everyone where the line was. It was not difficult to see where the farms stopped and the Park began, though: one side was dense tropical rainforest with giant buttress-trunked emergent trees, and on the other were the low cocoa trees. Signs warned anyone approaching the forest edge that entry was forbidden, as was hunting, fishing and taking wood.

Lieutenant Roy had brought one of his security detail with us and carried a machine gun. I was told the weapon was to guard against attack from wild animals, but as we walked by the boundary our protection scanned the farmland rather than the forest. We found a gap in the dense vegetation at the edge of the Park and went inside. It was a different world, darker and cooler, and dripping with humidity. There were fewer palms than I'd seen in the American rainforests, and as I looked up into the canopy I noticed there were plenty of ferns and mosses but no bromeliads—reminders that on different continents the character of tropical rainforests varies.

Although there have been incursions by illegal loggers, and cocoa growers have planted inside and under the tall forest trees on the eastern side of the Park, the authorities have, against the odds, largely held that line. But, despite the evident dedication and effectiveness of the team protecting the Tai Forest, the small resources they had to do their job was clearly not equal to the ever more challenging job at hand. In the end the rising tide of human need that surrounded the forest would become overwhelming.

Raiders of the rainforest

Across the forests of West and Central Africa it is not only the pressures arising from natural resource extraction, population growth and poverty that have eaten away at the forests and their wildlife, but also the deeply corrosive effects of corruption. It has oiled the cogs of ecological degradation and been a multiplier of the continent's social challenges, explaining why, despite the continent's spectacular wealth of natural resources, the majority of Africans remain impoverished. One place where I found out something about this was in the West African country lying to the east of Ivory Coast: Ghana.

I first visited Ghana in late 1990 with Friends of the Earth, along with my rainforest campaign colleague Simon Counsell (now the long-serving director of the Rainforest Foundation). We'd been tasked by Andrew Lees, our campaigns director, with the job of gathering information about illegal activities in the timber industry there. The plan was to investigate corrupt practices and produce a report that would be used to seek action from companies, governments and international aid agencies.

Our plan was to make contact with a secretive unit established by the provisional government led by Flight Lieutenant Jerry Rawlings. This was called the National Investigations

Committee (NIC) and its remit was to unearth and stamp out corrupt activities in the country, including its forestry sector. For a poor nation with limited opportunities to achieve economic growth, the loss of its forests to corruption was a major bar to improving people's lives. If money from the sale of logs was going into foreign bank accounts, then the exploitation of the nation's natural resources, whether sustainable or not, would contribute nothing to national development. Rawlings had set his mind to doing something about it. Lees had met one of his people on a trip to Ghana a few months before; hence the discreet invitation to Friends of the Earth to cooperate in shining a spotlight on what had gone on.

As dawn broke on a clear November morning, our Ghana Airways DC-10 began its descent over the southern fringes of the Sahara Desert. It flew on over a huge reservoir held back by the dam on the Volta river and over a belt of coastal rainforest to the Ghanaian capital, Accra. From the airport we were met by our contacts and driven off in an old Land Rover through leafy green suburbs and into the countryside. We bumped along dirt tracks, weaving between herds of goats and cattle, before arriving at a quiet location where some wooden huts were set beneath the shade of tall trees. A dozen men sat or stood outside between the buildings, most of them holding a rifle or machine gun—which said something important about the context of the Committee's investigations and how not everyone in the country welcomed the attention being paid to the rainforests.

We were ushered into an office and given a briefing about the work of the Committee and the fate of the country's forests. For years, we were told, forest trees had been cut without the correct permits and the most valuable timber smuggled out of the country with falsified papers. Some of the most expensive timbers had been declared as more common ones, denying the

Emergent rainforest trees – a rare sight in West Africa. Tai Forest National Park, Ivory Coast (TJ)

country the correct tax revenues. There had been some attempt to do more wood processing in the country, so as to create jobs and capture more economic benefit, but clapped-out old equipment had been sold to Ghanaian firms at well above the market rate by foreign companies. Having set out the broad picture, the Committee official turned to a cabinet and took out a thick bundle of folders. It was a dossier of evidence showing what had happened and who was responsible. He put it on the table and pushed it across to us.

We took the bundle and headed back to town before spending a couple more days meeting people working for rainforest conservation in the country, including colleagues at our excellent sister group, Friends of the Earth Ghana. We also traveled to neighboring Ivory Coast. Then we took the precious papers back to London, where our colleague Tim Rice set to work on a year-long forensic examination of letters of credit, invoices and tax papers to piece together how foreign interests had colluded with corrupt officials to defraud Ghana out of millions of dollars. As well as undermining development by stealing revenues that should have helped to provide clean water, equip schools and pay for the drugs needed by clinics, at least 250 square kilometers of tropical rainforest had been wrecked.

This was no isolated case. Almost wherever we looked a similar pattern appeared, raising big questions about 'governance': the quality of public administration, the reliability of the processes and the extent to which they are legitimate, trustworthy and truly reflective of the needs of citizens, rather than vested interests. So often, when looking into what's needed to save the rainforests, the matter of 'good governance' is cited. It can seem like quite an abstract idea, until you get down to what happens when it is lacking. In almost every case where the rule of legitimate public bodies is unreliable, corrupted or dysfunctional, then the rainforests have been degraded and destroyed.

Although we found it difficult to get serious attention on these subjects, our research into what happened in Ghana did help. When we finished our report about corruption in the forest sector in that country, we took it to producers at *World in Action*, at the time one of the biggest current affairs programs on British television. They used it to make a documentary called *Raiders of the Rainforest*, revealing not only the specifics of the corrupt practices and the companies involved, but also how the situation had been in large part created through the influence of powerful international organizations.

Structural adjustment

In Ghana the plunder of the forests had come on the back of economic crisis and international political pressure to open the country's economy to global markets. This was in the context of debts incurred during the 1970s. In the early part of that decade a glut of cash had built up in the global economy (in large part from oil price hikes) and led to loans being made to countries at low interest rates. With inflation at the time higher than interest rates, it looked like free money. However, in the late 1970s interest rates went up, while the value of commodities dropped.

Ghana was one of the countries that had run up huge debts, and when the crisis struck it found they were unrepayable. It was not alone. Right across Africa, Latin America and Asia, countries found that their debt repayment obligations could not be honored, leading to what became known as the Third World debt crisis. In a central position managing this huge problem was the International Monetary Fund (IMF). Alongside private banks and major Western governments who'd lent the money, the IMF set out a plan. It came down to establishing more relaxed repayment schedules for debtor countries in return for accepting conditions that opened up their economies so that they could export more, as a way to generate revenues to repay creditors.

It was called structural adjustment, and was a formula designed to let foreign companies in and exports out. It certainly succeeded in doing that in Ghana, but in the process it supercharged the corruption that helped destroy the country's rainforests. The economic reform package had been funded by, among others, Britain's Overseas Development Administration (now the Department for International Development) and the World Bank—a sister agency of the IMF. Companies rushed in behind the 'aid' money and the great forest rip-off got underway.

Alongside measures to boost exports, another demand of the IMF was to cut public spending, including that for conservation. One person who saw the effect of that was Professor Neil Burgess, the Head of Science at the UN Environment's World Conservation Monitoring Centre, who we met in Chapter 4. During the course of his work to protect Tanzania's exceptionally diverse montane forests, Burgess saw how attempts to address the country's debt crisis caused conservation to be put into reverse. 'They eviscerated all forestry departments of staff, which they deemed to be overstaffed. But these were people who were in the forest, doing guard duties and stuff like that. They were all let go because they had to reduce the size of the government, and when they did that there was nobody left out there.'

With the forests stripped of day-to-day protection, they became more vulnerable to encroachment by the rising population of poor rural people seeking land upon which to grow crops and to secure a livelihood. As the forests were cleared, so the unique wildlife they supported was diminished, an unknown number of species crossing the line to oblivion before even being described.

Even in countries with stronger governments, better enforcement and lower levels of corruption, structural adjustment programs added momentum for the exploitation of natural resources and thus deforestation. Even when it provided some short-term economic return, it became increasingly clear that these policies came with a hefty price tag.

14 CLIMATE AND COCOA CHALLENGES

Conservation of the rainforests is far more likely to succeed when smallholders are empowered to do sustainable farming

The large-scale loss of tropical rainforest can have dire impacts for local farmers—including the hundreds of millions of smallholders who, despite their poverty, produce more than half of the world's food calories. This was brought home to me in 2016, when I visited Ivory Coast and traveled to the village of Asabliko, about 20 kilometers down a dusty mud track from Méagui. I had gone there to speak with farmers who, as well as food crops, produced the country's biggest export commodity, cocoa. Tens of thousands of Ivory Coast farmers are employed in cocoa production and the country as a whole produces some 1.65 million tonnes per year—about a third of global supply.

Asabliko was surrounded by little plantations of cocoa trees, as was the rest of the landscape for miles around. The squat trees, cultivated to about 5 meters tall, were tended by farmers with just a couple of hectares each. The big pods bearing cocoa beans don't grow on twigs like most other fruits on trees, but directly from the trunks and large branches. As we walked between the trees one of the farmers explained that when the pods turn yellow they are ready for harvest and that each yields about forty beans, enough to make 50 grams of dark chocolate. Between the village huts, grass mats lay all about in the hot midday sun, covered with the reddish-brown beans being dried ready for bagging and shipping. Meantime, smoldering heaps of pods were being converted into charcoal to cook with.

As Ivory Coast's population grew, so demand for land upon which to produce this vital cash crop intensified. Asabliko lay about 5 kilometers from the border of the Tai Forest National Park and the community is one of several hundred that will determine whether that last large area of intact rainforest will survive. Cocoa trees are tropical rainforest plants and their

Cocoa being dried before bagging up and shipping with the help of a training program sponsored by Cocoa Life and run by US NGO Care (TJ)

natural habitat is in the understory beneath a canopy of taller trees. Planting in the last areas of rainforest was thus quite an attractive prospect for land-hungry farmers.

Today cocoa is grown right around the world in the wet tropical regions close to the equator, with conditions in the rainforest belt of West Africa being particularly suitable—at least until quite recently. For over the past couple of decades the climate has been changing, with hot, bone-dry winds coming from the Sahara, unlike those which normally came from the sea bringing rains that are recycled back into the atmosphere by the dense coastal rainforests. This is not good for cocoa production.

I sat beneath the shade of a small wooden communal building with a group of twenty farmers. Coucals, hornbills and weaver birds chirped, cooed and chinked from nearby scrubby forest and big shady trees in the village. But the sense of rural tranquility was lost as soon as the farmers told me their stories. The area had been hit by drought of unprecedented severity and the soil fertility had seriously declined. Yields had been poor and had been hampered, one of the farmers added, by varieties of cocoa trees that were susceptible to disease. The farmers were all in agreement, though, that the main problem was the erratic rainfall. For the first time ever it hadn't rained for four months. No one had seen that before, including the village chief, who was by far the oldest person there. Not only had this hit the cocoa harvest, but also the food the people grew to eat in their gardens around the village.

The farmers lived in a highly vulnerable situation. There was no river nearby from which to draw water for irrigation, and even if there was there wouldn't be money to build the canals and pumps needed to harness it. Like hundreds of millions of smallholder farmers across the tropics, and especially in Africa, they relied on rain. The more erratic that became, as it was becoming across so much of the continent, the more precarious became their position.

I asked why they thought it had become so dry and they answered almost as one: 'Because the forests have been cut down.' They knew, too, that it was the expansion of farming that had caused the deforestation. I asked what could be done. 'Plant more trees,' came one reply; 'put the forests back,' added another. The farmers said that if they didn't see improvements they wouldn't be able to make a living and they'd have to leave, to seek work in Abidjan and other cities. Some of the younger ones might be tempted to head north and risk the journey across the Sahara toward Libya and from there to Europe across the Mediterranean.

One of the cocoa growers, Amani Kwasi Unzui, a man in his mid-fifties, told me that he'd been in the cocoa business for thirty-nine years. He said that he could once make a reasonable living, but now it was a real struggle. He had an unusually large holding, a third of a square kilometer of cocoa trees, but even with such a comparatively large farm he'd struggled to survive. He explained that when he'd started out as a cocoa farmer there was much more forest and it often rained three times a day. No longer. I asked why the forest had gone. 'It's because there is more farming, because there are more people.' As the cocoa farmers talked of the dry winds coming from the north and landward direction, how droughts were the result, and the impact of that on their livelihoods, it was hard not to reflect on how their local wisdom was being corroborated by new science on the role of rainforests in generating rain.

All the major rainforest regions are different in how their progressive clearance might affect regional and wider rainfall regimes, but climate modeling suggests that deforesting West Africa (a process largely complete in Ivory Coast, Ghana and

Nigeria) could reduce regional rainfall by 40 to 50 percent. With that in mind, one might expect that at the center of discussions about how to sustain food supply in face of rapid global change might be strategies to stop deforestation. As famines, drought-driven shortages and higher food prices have during the last decade led to more widespread hunger for the poorest people, the question of how to feed the 9-billion-plus global population expected by 2050 has been increasingly on the agenda. But, while the UN Food and Agriculture Organization (FAO) has estimated that food output needs to rise by an estimated 70 percent by 2050, compared with 2005–7, little attention has been devoted to how such a vast increase in output must embrace a strategy that conserves and restores forests. It's a strategy that would be wise to pursue in relation to energy security too.

Power to the people

Near to the town of Soubré, some 80 kilometers from where I met the cocoa farmers, I saw another sector of the Ivory Coast economy that depended on rain. New pylons and cables strode out across the country from a vast construction site. They were connected to a new electricity substation that would soon be taking power from four new turbines embedded in the wall of Soubré hydropower dam. The turbines would be rotated by the awesome force of water running through the Sassandra river and for decades hence (if it kept on raining) would generate not only electricity but also a financial return on the $500 million dollar investment by Sinohydro—China's national hydroelectricity company. The dam would add more than 10 percent additional capacity to Ivory Coast's total electricity generating capacity and further dams were on the drawing board, as part of a plan to increase access to power across the country and end reliance on electricity imports from neighboring countries.

Soubré hydropower dam under construction on the Sassandra river,
Ivory Coast, June 2016 (TJ)

Walking onto the construction site I found a Chinese
engineer who seemed like he might be in charge. I asked him
if I might walk around to get a sense of the project. He was
covered in sweat and evidently a little flustered as trucks and
bulldozers rushed about in all directions, kicking up clouds of
reddish dust. He managed not only a friendly nod of consent
but also a proud welcoming smile and summoned one of his
Ivorian staff to show me around. They explained how the dam,
just above natural falls and rapids that caused the river to drop,
took advantage of the local geology to generate an output of
270 million watts. It would also create a 50-square-kilometer
reservoir.

Large hydropower dams are complex in terms of economic viability, the impact they cause on people as they flood land, and what they do to rivers and wildlife, but this one was nearly done. Within a year or so it would be producing a high proportion of Ivory Coast's electricity. For how long it would do so was another question. Many hydropower dams have suffered from the effects of siltation caused by upstream soil erosion, which is in turn the result of deforestation. The eroded material fills dam reservoirs, reducing the amount of water it holds and thus cutting power output. The color of the river showed that this one was evidently at risk from that, but also falling river flow arising from reduced rainfall, also caused by deforestation.

Every stream and river on Earth only exists because the water that flows through it was first lifted into the air and transported by clouds and then rain to the headwaters from which it originates. Evaporation from trees is a vital link in the cycle that makes this possible. When the water elevator falters, power cuts can follow. A while before my visit to the dam I had talked with a young Ghanaian woman who told me how her company had been forced to lay people off because they had been hit by repeated power cuts. The reason for this was a drought that had hit the Akasombo Dam on the Volta river. This huge piece of infrastructure was built in the early 1960s and became vital for the economy not only of Ghana but also neighboring Togo and Benin. Lake Volta, above the dam, is fed by the Volta river and has the largest surface area of any man-made freshwater reservoir in the world, covering some 8,502 square kilometers. That water provides the means to generate more than 1,000 million watts of power—when it has rained enough. But in the more than half-century that the dam has been in operation, it has been subject to a number of droughts, and as the forests have been cleared, so the threat to its economic viability has grown.

In southern Africa, too, the effects of drought have hit the power sector. From its source on the southern fringes of the Congo Basin rainforests, the Zambezi river flows downstream to Victoria Falls, and from there to Lake Kariba on the border of Zambia and Zimbabwe. That vast water body was formed in 1958, when the sluices on the newly built Kariba Dam were closed, leading a few years later to what was then (by volume) the largest man-made lake on Earth. The dam and the electricity it generated made Zambia something of an African success story, harnessing power for economic development and poverty reduction in ways that few other countries on the continent matched.

Ghana's vital Akasombo Dam generates electricity by harnessing the rain that flows in the Volta river (ZSM/WikiCommons)

During 2016 and 2017, however, a punishing drought caused power outages that led Zambia's main export sector—mining—to cut back production. This in turn increased the mining companies' costs, cutting their competitiveness in global markets. Exports shrank, leading to lay-offs, reduced tax revenues and inflation. The extent to which that specific drought was linked with climate change or deforestation is not clear. What we do know, however, is that the links between rainforests, rainfall and climate are fundamental and that reduced forest cover can have serious ramifications for water security, meaning that as the forests decline, so economic and social risks increase.

Scientists looking at how environmental change might pan out in the future increasingly see the need to look not only at individual trends, but also the connections between them. One manifestation has been a growing emphasis on what has been described as 'the food, water and energy nexus'—that is, the set of linkages between these vital areas of human need. Some leading thinkers, including a former British Government Chief Scientific Adviser, have warned of a 'perfect storm' arising from the collision of rising demand for all three at once, which is what is happening as a result of economic expansion and rising human population. The clearance of the tropical rainforests is fueling that storm and determining its severity and impacts.

Key to balancing the equation is the well-being of poor rural populations. Over time it has become ever clearer that if they can't sustain themselves or their families, then they'd be unlikely partners in sustaining the forests. This is today one of the main challenges—finding ways to help rural people while at the same time avoiding forest clearance. To do this, a new style of development would be needed, one that lifts incomes while keeping the last rainforests intact—and indeed expanding them.

Helping the cocoa communities

When I talked to officials at the Tai Forest National Park in Ivory Coast they told me that its successful preservation was not so much about enforcing that boundary that I'd walked along with armed security, but more livelihoods and education. These, they felt, were key to holding the rainforest front line. I recalled research on the relationship between female literacy and family size, and how data showed that when women were educated, they tended to have fewer children and better incomes. Out there by the forest, where people lived in grinding poverty and whose large families led to increasing demand for farmland, it seemed like a very logical priority. I was encouraged to find that this was the aspiration not just of the man responsible for a National Park but for the world's largest chocolate company: Mondelēz, whose many brands include Cadbury's, Côte d'Or and Toblerone.

In 2012 Mondelēz launched an initiative called Cocoa Life, through which it would invest some $400 million in supporting 200,000 of its cocoa farmers across the tropical rainforest countries of Ghana, Indonesia, India, the Dominican Republic, Brazil and Ivory Coast. In Ivory Coast $100 million was allocated to assist 75,000 people in cocoa-growing communities. In addition to support for education, farmers were helped to improve yields and earn higher incomes, while at the same time sustaining the landscape in which the cocoa was grown, including the remaining rainforests. The idea was to strengthen communities and thus make cocoa a more attractive business for young people. It was an attempt to implement a truly integrated agenda and there was a clear business case for doing it, even for a profit-driven multinational.

The company had realized that if in the years ahead there were fewer farmers wanting to grow cocoa, then the company's chocolate business would be at risk. It would also face uncertainty and rising costs if deforestation led to the kind of long droughts that had recently caused yields to fall.

As well as major companies, the Ivory Coast government had an interest, for if cocoa declined as an export crop and a source of employment, then civil unrest could follow. Adoulaye told me that he was confident that the deeper and more fundamental pressures on the forests could be brought under control if only he could succeed in showing people the Park was for them, and not against them. He told me how, when he asked communities to join in with the conservation of the Park, they asked him, 'What do we get?' One of the things he tells them they get, and which I found the cocoa farmers agreed with, was that the Park produces rain, and thus cocoa and income.

As I listened to Adoulaye, it was clear how the multinational company and National Park were actually trying to do the same thing: improve the well-being of the farmers. The Park gets its forest better protected and the company gets a more secure cocoa supply chain that is dependent on not only motivated and skillful growers, but also rain. Economy, people and forest all at once.

Raising incomes and shade

I met members of ICOM, an Ivorian cocoa cooperative, at the town of Méagui, close by where I had discussed drought with the farmers. The co-op told me how they had been receiving practical help from a partnership between Care, a US-based NGO, and the Mondelēz Cocoa Life program. A crucial aspect of the plan involved buying pledges from Mondelēz so that growers knew how much cocoa they would sell at harvest time. Orders were made three years in advance, so long as suppliers met the standards of the Cocoa Life program.

We talked in a warehouse run by the cooperative, along with workers from Care. It was the place where cocoa was collected after harvest and received the produce from over 1,000 small-scale growers before it was trucked to the port of San Sebastian. ICOM expected to collect about 3,000 tonnes of cocoa from a combined area of about 6,000 hectares (60 square kilometers), which meant each hectare produced on average 500 kilograms. At the government fixed price of $2 a kilo, a smallholder with two hectares would thus gross about $2,000.

From that income, costs had to be deducted, including about $350 for fertilizer. A huge amount of labor was involved in generating that return and it needed to support entire families. Even assuming no other farm costs, for a family of four (and most farmers have more than two children) that put the per capita income at considerably less than $2 per day—the current threshold measure of absolute poverty. No wonder many growers wanted to look for a livelihood elsewhere, or seek to increase their income by encroaching into the forest. In the wake of another drought, more would try to get on a boat to Europe.

Although education was a necessary step for these communities, it would take some years for it to deliver benefits. The farmers needed additional practical help right away. That came in part through the supply of new higher-yielding cocoa trees. ICOM had 3 square kilometers of nurseries growing tree saplings for farmers. With better trees and training on how best to keep them, they might in a few years get up to 2 tonnes per hectare—a four-fold increase in yield, and thus a four-fold increase in income (so long as price guarantees remained and sufficient rain fell).

In the nurseries ICOM were also growing saplings of native rainforest trees for distribution to the cocoa growers. As they grow and mature in the fields, with the smaller cocoa trees beneath, these form a leafy canopy that shades the crop. The wild cocoa trees that the Maya Indians found in the rainforests of Central America was a species of the rainforest understory and rearing its cultivated descendants beneath a canopy that mimics the forest brings several benefits. This includes retention of moisture, making the most of the rain that does fall and increasing the resilience of the crop during droughts. The tall trees also protect soils during heavy rains and reduce nutrient loss, with nutrient recycling improved as the fall of leaves return organic matter to the soil surface.

The shade trees can also help suppress aggressive weeds that thrive in open and lighter conditions. And the big trees also lock up carbon and provide a habitat for native birds that in turn help with natural pest control. Best practice cocoa planting under shade was also using agroforestry methods to combine cocoa with bananas and legumes. The farmers were broadly positive, as they could see the benefits. Community group representatives were also taking an active role in discouraging illegal logging in the remaining forests. Being sanctioned by the government or the National Park authority was one thing, but having your neighbors urge you to protect the forest was evidently something else. And in the years ahead, when farmers harvest legal timber from the shade trees on their own land, they won't want to be undercut by illegal supplies.

The fact that all this was being supported by a multinational was an interesting turn of events, as not too many years ago such companies would simply point to market forces and short-term profitability as the drivers of their business strategies. But you don't have to buy the idea of corporate social responsibility to see why Mondelēz was involved. Self-interest also dictated this approach, when risks to the company's profits were visible in failing soils, reduced rain and worker dissatisfaction.

The fact that other cocoa giants, including Nestlé and Mars, have launched comparable cocoa plans underlines the extent to which the message is getting through. In 2017 the world's largest cocoa companies came together at a meeting in London hosted by the Prince of Wales to make a joint pledge to end deforestation in the landscapes from where they source their key ingredient. If they stick to that they could during the years ahead make a huge difference, helping many more smallholder farmers switch to agroforestry-based approaches—the kind of indigenous farming that in some ways mimics the natural ecology of the rainforest, with its many layers, nutrient recycling and hydrological characteristics.

Community development

I traveled through the cocoa landscape of western Ivory Coast with a Mondelēz employee called Mohamed Amin. He grew up in the country and knew its recent history and current challenges. He explained the business case for the protection of the forest and how a longer-term view was needed than one simply responding to market supply and demand: 'We need to take action now to make sure we can get anything in the future. We need to keep the forest and we need to expand it.' He told me this in front of a group of farmers. Judging by the nods of agreement it was a message that was beginning to get through.

I asked the farmers themselves if they felt the Cocoa Life program was helping them. 'Yes,' several of them said emphatically. One added that the most important thing was learning how to get more from their little plots of land. A Care worker told us how his job was to support eighty-one farmers, going from farm to farm. Mondelēz supplies him with a motorbike and petrol, and his job is to impart knowledge and advice while at the same time ensuring the standards of the Cocoa Life plan are being met. His work enables a document chain to be created so that, for example, a Toblerone in an airport shop in California, Tokyo, London or anywhere else can be traced back to the farmers that supplied the cocoa. This enables the company producing it, and the person eating it, to know that it was produced as part of a program to improve the livelihoods of the people while saving the forests they live alongside.

Farming advice was one thing but community development was even more challenging. The closest primary school to that village where I sat in the shade with the farmers was 3 kilometers along a dangerous track. Basic medical facilities were the same distance and the secondary school 9 kilometers distant. Most of the children thus didn't attend school. The village also lacked sufficient fresh water and during the prolonged recent drought they had only a single handpump with which to tap groundwater—for 1,062 people.

Adrien Konano, a Care program trainer, told the farmers how another community had invested their profits from cocoa into making a fish farm and that this was making them money and lifting living standards. 'When we know the problems that lead to poverty, we can solve them,' he explained. 'When you use your head as well as your hands, there's no reason to be poor.' With that, they all rose to their feet and declared that they all wished to be rich—which, it seemed, meant having clean water, a corrugated metal roof over their head, schooling for the children, basic healthcare and mattresses to sleep on. Amin added that if they made more money, they could buy meat and could stop hunting the local wildlife to extinction.

I looked at the farmers' faces. They seemed to agree, but I could understand the hesitation they might feel in rising to the challenges they faced. I asked one cocoa farmer what he'd like to say to people in Europe, the USA and Canada, where people enjoyed chocolate. 'I'd like to thank them,' he said. 'We don't see

Mondelēz's Cocoa Life program is one among several initiatives that seek to improve the livelihoods of smallholder farmers (Cocoa Life)

the people who eat the chocolate, but Cocoa Life is helping us to do better.' I felt very humble and responsible for the choices I made as a consumer. After all, whatever you buy in a shop— margarine with palm oil in it, chocolate, coffee, meat and dairy products produced with soya, wooden furniture and paper—all of them shape outcomes on the ground, one way or another.

The business case

Traveling back to Abidjan I passed through the huge shanty towns that had sprung up around it. Many of the people living there were former farmers who'd had enough of trying to make a living from the land. Whereas they once threatened the remaining rainforest, now they were putting a massive strain on inadequate urban infrastructure. It was a pattern repeated across the tropical world, as population growth and rural decline, in part caused by deforestation, were driving utterly unsustainable urban expansion.

In the city I met Mbalo Ndiaye, who runs the Mondelēz operation—and Cocoa Life—in Ivory Coast. He told me how Mondelēz had entered a new phase, going beyond labels that said cocoa was more or less well produced. 'Our plan goes beyond certification to make direct contact with the farmers. The aim is to improve the social conditions of the people. It's a whole package. Certification is farmer-focused but we also need an area focus, to look at the whole landscape. Farmers have their own development plan and we work with NGOs to build the community capacity.'

He told me how the company had put in place a long-term plan that was costly but made perfectly good business sense. 'We need to take a landscape view and community development view; otherwise people will leave and move to the cities. It's especially important to help the young people. They are the future.' Ndiaye pointed out how the plan included targeting resources around the Nawa district because of its proximity to the Tai Forest, with its forest elephants and chimpanzees. 'We are trying to build collaboration in the buffer zone to work with the farmers there, working with the National Park and helping them with resources.'

I asked him if it might be possible to sidestep some of the complexities of dealing with multiple small farmers by producing cocoa on an industrial scale, like the palm oil industry had done in Asia. He told me it could be done, but that the social downsides would be devastating, especially considering the shortage of jobs and livelihoods in the countryside. As well as sustaining jobs, Mbalo highlighted the role of education: 'We need to help the farmers keep up with technical developments. That is difficult when about 40 percent of them are illiterate. It's not sustainable. We are investing in primary education to help them prepare the next generation and also to prevent the child labor problem. Many farmers are presently illiterate, but their children needn't be.'

The role of government

It was encouraging to hear this perspective from a major global company, but having spent so much of my time over the years advocating that governments do more to protect the forests, I was interested to see what was on offer from that quarter. To find out I went to see Jean-Paul Aka, a young civil servant working Ivory Coast's national government REDD office in Abidjan. REDD, of which more later, stands for 'Reducing Emissions from Deforestation and Degradation' and is a United Nations initiative to protect rainforest.

I've visited the offices of many environment ministries over the years and often had an immediate sense that they were at the bottom of the political pecking order, with clapped-out facilities and few resources. But the office looking after the conservation and restoration of the Ivory Coast's forests was promising, housed in a nice building, with modern facilities and a big new sign outside.

I was welcomed by Jean-Paul, who told me his work was to devise strategies that could harness the growing pot of international finance for the conservation and recovery of tropical forests. He said that, since independence, the country had prioritized agriculture as a means to achieve growth and development for its fast expanding and mostly very poor population. But that was at the expense of the forests. He explained the complexities that come with the need to boost exports while keeping the forests. 'We're working with companies to help smallholders produce more on the same land, while getting more trees in the farmed landscapes, including native fruit trees that can also produce food and income.' He understood, too, that the forests were an economic issue: 'The loss of the forests is affecting the rains and that is affecting the cocoa production. Around the Tai Forest the rainfall is very good, and that is one reason why we need to keep it there. This is why we see big potential to reforest the cocoa landscape, including shade trees, to help increase production while taking in carbon and thus fighting climate change.'

Jean-Paul added that it was vital to have the cooperation of the private sector: 'They have an interest in this, not only because it will help them to produce what they need but also to reduce the risks that come with their products being linked with forest loss. If we work together at the scale of the landscapes, we can achieve economic and social advantages for everyone.' In making such partnerships work, the government had received money from the World Bank and French overseas development ministry, some of which was being used to leverage cooperation and resources from companies. A growing priority for the funding agencies was fighting climate change and the plan was to develop longer-term strategies and to apply for more international resources for that purpose.

Jean-Paul estimated that a $70 million grant would enable the country to cut carbon dioxide emissions by a third by 2020. The plan was to target several areas: one around the Tai Forest National Park, one in the southeast toward the border with Ghana, one in the center of the country, and another in the north. The idea in the northern zone was to encourage the large-scale restoration of natural forests to increase the regional rainfall and thus improve conditions for cocoa. Another benefit that would come from restoring mosaics of natural forests among the farmed areas would be disease control. Swollen shoot disease is caused by a virus and can devastate cocoa yields. An outbreak in the southwest of the country that began in 2003 reduced output in some areas by two thirds. Areas of forests lying between plantations would limit its ability to spread.

To do all this, three broad plans were being worked up: one for protecting remaining natural forests, another for restoring them, and one to promote sustainable farming. Aka's team had estimated that conserving natural forests kept 300–400 tonnes of carbon per hectare (10,000 square kilometers) out of the atmosphere, forest restoration took out 250–350 tonnes, while agroforestry (those shade trees over the cocoa) would soak up about 150 tonnes per hectare. He was still working on the figures but told me his provisional estimate that it would cost about $400 per hectare to conserve natural forest for ten years, about $1,600 per hectare for reforestation for five years and about $1 per tree planted in farmed landscapes as part of agroforestry programs.

All these numbers will be subject to change, Jean-Paul said, but the point was that there was a real economic value to protecting and restoring the forests and that real money can be spent on making it happen. He said that his country was ready and willing to help realize global climate benefits and pleased to be paid for that service, but made the point that the forests were not for sale. They were Ivory Coast's rainforests and, while the country would welcome help to conserve and restore them, the forests would still belong to them.

As we began to wind up our meeting, I asked him if he was hopeful. 'I'm very optimistic,' he replied. 'We need many more optimists, however, and for them to assist us in the conservation and restoration of the forests.'

It is not easy to foster that optimism when faced by Africa's challenges—poverty, a rapidly rising population, pressure on natural resources, economic crisis and corruption. All these factors have contributed to widescale forest loss and degradation across the moist tropical regions of Africa. New approaches toward integrated development are encouraging. But there are new threats, too, not least the specter of large-scale commodity agriculture, in particular for palm oil, which will become an increasingly powerful pressure in Africa, as it has in the rainforest regions of Asia.

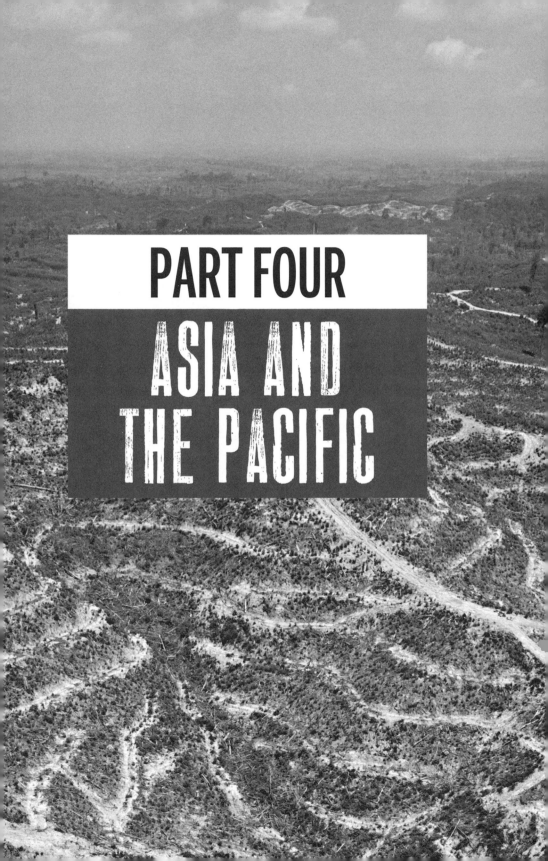

PART FOUR

ASIA AND THE PACIFIC

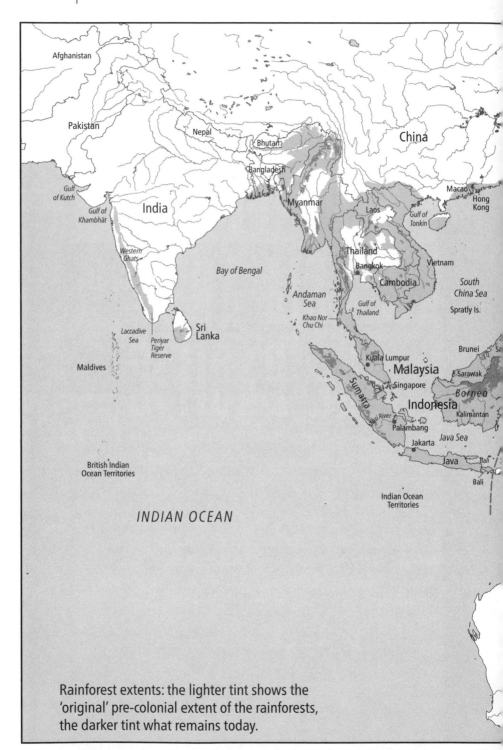

Rainforest extents: the lighter tint shows the 'original' pre-colonial extent of the rainforests, the darker tint what remains today.

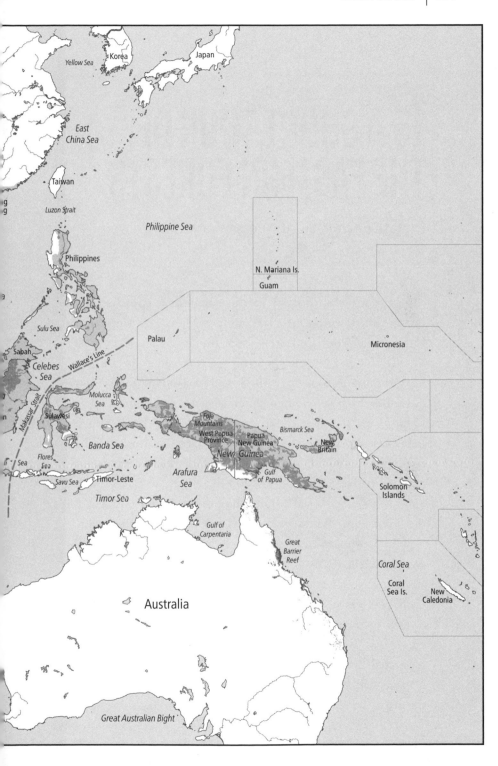

15 A SHORT TOUR OF THE EASTERN FORESTS

Explosive economic and population growth have taken a heavy toll on Asian and Pacific rainforests

The rainforests of the Asia Pacific region once stretched from India through Burma, Thailand and the Malay Peninsula, across Indonesia and out to Australia, the Pacific islands and Hawaii. Much of the original forest is now gone. The Indian subcontinent has lost most of its primary forest, and lowland rainforest is now cleared or at best fragmented in Malaysia, Thailand and the Philippines. The most significant Asian rainforest that remains today lies across the Indonesian archipelago—and it is seriously under threat. The island of New Guinea has the Pacific region's most extensive intact forests, but there, too, the demand for natural resources is set to raise the pressure—and not only on the forests, but also the indigenous people living in them.

As in South America and Africa, advanced societies existed in Asian and Pacific rainforest for thousands of years before their 'discovery' and exploitation by Europeans. A recent study in Cambodia using advanced airborne laser scanning technology revealed numerous 'lost cities' of the Khmer empire beneath the rainforests. These were extensive formally planned urban areas supported by farmed landscapes, quarries and large-scale water management systems. The Khmer capital, Angkor, was during the early part of Europe's medieval period the world's largest city, covering an area larger than modern Paris. The Khmer empire stretched across much of mainland Southeast Asia, with far-flung settlements connected by a network of highways, before its collapse in the fifteenth century, possibly due to an overextended urban population supported by an insufficient number of people producing food. Indonesia, too, was in medieval times an important trading zone. Marco Polo, traveling through Asia in the 1270s, found the archipelago 'producing pepper, nutmegs, spikenard, galingale, cubebs and cloves and all the precious spices, visited by great numbers of ships and merchants.' He described how people he came across in Sumatra used to be Hindus but had, following contact with Muslim merchants, recently converted to Islam.

Compared with the societies that Columbus encountered in the Americas, Polo came across more outward-looking people. Around the Asian coasts were connected societies with a history of long-distance trade and cultural sophistication. Some were colonists themselves, from India and China.

Perhaps this, and a degree of resistance to Old World diseases, gave them a level of resilience to the European exploitation, which to begin with was relatively light. One reason for that was the difficulty in finding the correct winds needed to round the southern cape of Africa and enter the Indian Ocean, which meant that Europe's initial colonial activity was directed across the Atlantic. It was not until the sixteenth century that Portuguese and then Dutch ships began to visit Asian territories in search of spices. They were followed by British and French vessels, which also ventured out into the vastness of the Pacific.

Asian societies and cultures were drastically affected as the Europeans established their colonies, but their emphasis was largely on commerce, and local political structures often remained in place. That said, there was always the implicit threat of force and in some cases outright and brutal subjugation of some

The red junglefowl, the wild ancestor of the domestic chicken
(Francesco Veronesi/WikiCommons)

uncooperative rainforest societies; for example, the population of
the Spice Islands by the Dutch. The trade also created the world's
first transnational corporations: the British East India Company,
in 1600, and the Dutch East India Company two years later.
Preceding the rise in the power of modern corporations by
centuries, these Companies possessed the legal and practical
means to wage war, imprison, negotiate treaties, issue currency
and establish colonies. Once they became established, the
European colonial powers dominated the region for centuries:
the British had most of India, Ceylon, Burma, Malaya and
Singapore; the Dutch, Indonesia; France held Indochina; Spain,
the Philippines; and Portugal, Goa.

For Asian rainforests, the colonial period was essentially a story of exploitation and export of natural resources—spices, fruits and timber. Asia's rainforests impacted on the West, too, in the spread of a range of food sources, domesticated by indigenous peoples long before the Europeans arrived. The origins of some of those foods can be seen in surprising settings.

Asian forest foods

One afternoon, walking in a park in Singapore, I stopped in my tracks to look at a group of red junglefowl—a male bird with its flowing black and iridescent-green tail followed by two females and twenty speckled chicks. It was remarkable that such wild creatures could live among the skyscrapers and flyovers of this most urbanized rainforest country. And they were a reminder of just how much of our basic food has rainforest heritage, for these birds are the direct ancestors of the chicken.

First domesticated in China, these rainforest birds reached Europe by way of central Asia around BCE 3000. They were perhaps crossed at some stage with a similar species called the grey junglefowl, a bird that I've encountered in a very different setting in wild upland rainforests in western India. From little flocks in smallholders' gardens to vast factory farms, we now raise in excess of 50 billion of these birds each year. In 2014, approximately 179 domesticated junglefowl eggs were available for consumption by each person on Earth—a source of high-quality and inexpensive protein, selenium, vitamins D, B6, B12 and trace minerals including zinc, iron and copper.

Tea, the world's most popular drink, also has rainforest heritage, coming from a genus of evergreen shrubs that occur in moist upland forests in southwest China. Today, the shrubs' leaves are harvested from plantations across tropical rainforest countries including India, Vietnam, Indonesia and Sri Lanka.

The trade underpins millions of jobs, sustains the profits of huge multinationals and forms a substantial part of the tax revenues of many countries, as does the sugar that many people add to a brew.

Sugar cane was first domesticated by rainforest peoples, probably in New Guinea, about 8,000 years ago. The tall grasses from which sugar cane is descended grow wild in the hot wet tropics and along riverbanks and in clearings created by tree falls or animals. Their cultivation generated a supply of sweetness previously only available from scarcer sources, such as honey. Entrepreneurial European travelers saw that fortunes were to be made from supplying an eager public and sugar cane rapidly spread across the tropical rainforest belt. Columbus, on his second voyage to the New World, took sugar cane to Hispaniola and it was later grown in huge plantations on the Caribbean islands. It is cultivated today across the tropics, including in Thailand, Colombia and Indonesia, and in Brazil it is used increasingly for biofuel, offering a renewable alternative to petrol.

Many familiar fruits have tropical rainforest origin, too, particularly the vitamin-C-rich citrus ones used so widely in soft drinks and confectionery. Genetic research suggests that all citrus species (about thirty-five in all, including mandarins, grapefruits, kumquats and limes) have a common ancestor that evolved in the distant past in Australia. Lemons probably originated in the forests of Burma or southern China, and oranges from wild trees in Southeast Asia. Bananas are another Asia-Pacific rainforest fruit. Wild species are found throughout Southeast Asia but are mostly inedible. Occasionally, though, mutant plants bearing crescent-shaped fruits appear and it was from these that people first domesticated the species. The aberrations were seedless, so growers took suckers from the base of the plants to grow genetically identical new ones.

It seems this first occurred about 10,000 years ago in New Guinea. Now, of course, the banana is a massively valuable cash crop throughout the tropics and subtropics, supporting millions of farmers' livelihoods.

Rice, timber and palm oil

Another global staple is rice, the most widely grown tropical cereal. There are thousands of varieties today but they are all descendants of wild ancestors from the swampy banks of rainforest rivers in South and Southeast Asia.

It was the success of rice production in Asia that led to the first significant deforestation in the region. The extraction of spices, woods and other products had been achieved without major damage to rainforest ecosystems until the early twentieth century. Alfred Russel Wallace—who in 1858 came up with a theory of natural selection at the same time as Charles Darwin, based on his travels in Asia—wrote of the rainforest's seemingly limitless fertility, saying he was tempted 'to convert the virgin forest into green meadows and fertile plantations.' And that was exactly what happened over the next half-century as large areas of lowland forest across Burma, Thailand, Cambodia, Laos and Vietnam were cleared for rice production and export, as were parts of Indonesia, such as Java and Sumatra. By 1920 some 100,000 square kilometers had been cleared. In the Philippines 300 years of Spanish colonial occupation had by then also taken a heavy toll on the forests.

During the post-war years of the twentieth century the pace of forest loss quickened right across the region, spurred by international demand for timber. This was initially caused by the rapid economic growth of Japan, which by 1960 had become the world's largest importer of tropical timber. Japan, ironically, protected its own forests, and indeed has the highest

The fruits of the West African oil palm are a productive source of vegetable oil, with vast plantations supplying tens of millions of tonnes to world markets each year (WikiCommons)

forest area (60 percent of its land) of any industrial country outside Scandinavia). But its timber companies secured rights to exploit the rainforest trees of Borneo and the Philippines, for everything from hardwood furniture and paper to disposable chopsticks and building materials.

By contrast with Latin America and Africa, Asian forests held a high density of commercially valuable tree species, including dipterocarps and ramins. These hardwood trees, between them comprising around 100 different species, dominated the canopies of many Asian rainforests, with particular diversity in Borneo. Their high value meant that logging was more intense and more damaging than in other tropical regions, where the most desirable species were more scattered through the rainforests. Logging operations in Asian rainforests often take out half the mature trees, causing severe degradation of the forest, meaning that a second cut might not be valuable, even if the logged area is left for thirty or more years. This hastens total deforestation, with logged rainforests often completely converted after one cut of timber to plantations of oil palm. Across the region, but especially in Malaysia and Indonesia, the lucrative rewards of heavy logging were used to finance the establishment of palm oil plantations, on a massive scale. Where tracts of tropical rainforest once stood, with their diverse plant and animal life, there are now industrial monocultures of oil palms.

The West African oil palm was first established in Southeast Asia in the late nineteenth century and was an immediate success, due to the favorable climate and the absence of the pests and diseases that attacked it in its native Africa. Selective breeding and improved cultivation practices helped to increase yields dramatically, with a fruiting head on a high-producing palm able to produce hundreds of plum-sized capsules. It is an astonishingly high-yield crop. The yellow flowers of oil seed rape (canola) that cover swathes of British farmland typically produce about 3 tonnes of oil per hectare. A good oil palm plantation can get nearly triple that.

In 2016 almost 65 million tonnes of palm oil were harvested around the world, with 55 million tonnes coming from just two countries—Indonesia and Malaysia. We are all, however, part of the palm oil cycle. It is in vast numbers of consumer products— margarine, biscuits, ice cream and crisps, as well as shampoo, soap and cosmetics. It is often listed as 'vegetable oil' or 'vegetable fat,' and can be processed to yield a range of oils, from the very fine and valuable to cruder 'bottom of the barrel' grades. Demand has also been driven by the increased use of biofuels. In some parts of the world, such as the European Union, policies have been introduced to begin a shift from petrol and diesel made from fossil oil to embrace fuel made from plants, including palm oil (to make biodiesel). Sometimes presented as a 'green' policy, this has increased the rate and scale of plantation establishment, very often at the expense of the rainforests.

It is not only the outright replacement of forests with oil palms that has been ecologically harmful, but also the fragmentation of larger areas of forests. The plantations cause pollution impacts, too, as fertilizers and pesticides get into watercourses. Fertilizer use also causes emissions of the powerful greenhouse gas nitrous oxide, while forest conversion to oil palm is a major source of carbon dioxide. Oil palm plantations have a lower biomass than the natural rainforests they often replace, and therefore lock up less carbon. Even worse, many Asian oil palm plantations have been established on peat swamp forest and, as large areas of that ecosystem have been drained, burned and converted to oil palm across Borneo and Sumatra, they release carbon dioxide and cease to be systems that actively accumulate carbon.

Borneo: the Penan

The pace of rainforest damage driven by the timber to oil palm dynamic is most dramatically illustrated in Borneo, the world's third largest island. Satellite data reveals that the island's natural forest cover crashed from 76 percent in 1973 to 28 percent in 2010. In Sarawak and Sabah some 80 percent of the rainforest was severely degraded over just two decades, between 1990 and 2009. And when large areas of rainforests were depleted (as in the Americas), it was not only the forests that were casualties but also the indigenous forest peoples.

During the mid-1980s the Penan resorted to peaceful logging-road blockades to protect their lands (Bruno Manser/Global Witness)

The Penan people once lived throughout the forests of Sarawak but during the late 1980s and early 1990s found themselves on the front line of government-sanctioned timber extraction. They defended their lands in concert with Western campaigners (including our rainforest team at Friends of the Earth). First their forests were logged and then large areas converted to plantations of oil palms. It was part of an economic strategy that regarded the forest people as obstacles to 'progress' and a barrier to countries asserting their 'right to development.'

As the Penan's rainforest was destroyed, so too was their way of life. One campaigner especially focused on the plight of the Penan was the Swiss rainforest activist Bruno Manser. I met Manser on a few occasions, including the G7 Summit in London in 1991. At the end of that high-level political meeting, and as the closing press conference was being hosted by Prime Minister John Major, Manser climbed a tall lamppost outside the conference center and unfurled a banner calling on the world's most powerful leaders meeting there to use their influence to protect the rainforest home of the Penan people. It was one more intervention in what was for him already a very long campaign.

By then Manser had spent six years in the rainforests of Borneo, living with the Penan. He'd helped them organize blockades on logging roads so as to prevent timber companies getting access to the forest. His activities that day in Central London earned him a lot of front pages and headlines, adding to what was for the Malaysian government a most unwelcome focus on their domestic affairs. He was declared 'number one enemy of the state' and, when he went back to the forests and the Penan people, special units were sent in to look for him. He was arrested twice but on both occasions escaped. The Malaysian Prime Minister, Mahathir bin Mohamad, blamed Manser for disrupting law and order and wrote him an open letter saying that, 'As a Swiss living in the laps of luxury with the world's highest standard of living, it is the height of arrogance for you to advocate that the Penans live on maggots and monkeys in their miserable huts, subjected to all kinds of diseases.'

In 2000 Manser returned to the forests and, although he'd previously survived the hazards familiar enough to the people he worked with, this time he disappeared, never to be seen again. Although his body was never found, he was in 2005 declared dead. It was a tragic end for a charming and determined campaigner who made an outstanding contribution to the world's understanding of what was happening to the rainforests.

As well as the specific spotlight he shone on the cause of the Penan people, his campaign said a great deal about the divide that had emerged between those who saw forest peoples living in harmony with vital ecosystems, compared with those who saw them living in intolerably primitive conditions and blocking 'the right to development.'

There are about 10,000 Penan people left today, but because of the degradation and destruction of their forests perhaps no more than sixty families live a traditional, nomadic existence in the rainforest. As has been the case among other indigenous peoples, the loss of their ancestral ways has led to a variety of health problems—caused not just by changed physical conditions but also the loss of identity. The Penan see themselves as inseparable from the forests, and when that was taken from them, their inner world was lost too.

Transmigration

In Indonesia the state-sanctioned assault on the lands of indigenous societies was similarly determined, and proved equally destructive. There, it was caused not just by heavy logging and oil palm plantations but also by a mass resettlement program. From the late 1960s through the 1990s, the Indonesian government shipped out more than 3 million people from the densely populated central islands of Java and Bali to so-called outer islands—Borneo, Sumatra and Sulawesi. The transmigration policy, as it was called, led to an influx of people into some of the most biologically diverse and untouched rainforests on Earth.

The idea was to assist the prospects of poor farmers, but the consequences were dire, for, while the transmigration program made little difference to the population of Java and Bali, nor to poverty levels there, it impacted heavily on the sparse populations receiving the incomers. Many settlers were moved into areas already occupied by indigenous groups. As a result there were conflicts, sometimes involving violence.

It was a disaster not only for the original forest people, but the forests too. The clearance of land to make way for crops to feed the settlers and their families led to massive deforestation, and still does, as those migrants and their descendants continue to encroach into the remaining forest lands, often through illegal logging followed by burning to clear land and establish small oil palm plantations. And, despite opening new lands, many of the transmigrants have found it hard to make a living. As in the Americas, rainforest land in Asia has poor soil for growing crops, at least without major use of fertilizers. And even when families did manage to get decent yields, the lack of infrastructure in these remote areas meant poor access to market. So the scheme didn't cut poverty, it just moved it from one part of the country to another.

Much of this was foreseeable but even then the politics made some kind of sense. By moving people out of the politically important central islands, the leaders who promoted the program gained favor among the millions of people to whom they distributed free land. At the same time they could colonize the outer islands with a politically sympathetic population, helping to subdue potentially rebellious indigenous groups. They could also move a workforce to remote areas to assist with the future exploitation of natural resources.

In 2015 I met some of these transmigrants in Sumatra, in a village called Mulya Agung, located near a mangrove forest on the coast of Jambi province. The settlers had come from Java during the early 1990s and on arrival each family was granted two hectares (20,000 square meters) to make a living. On these little plots—a typical land allocation for the transmigrants—they and their families tend to grow a combination of oil palms, coconuts, vegetables and rice, and rear small cattle (domesticated from the wild banteng cattle indigenous to Bali and Java). About 1,500 people lived there and, although they still spoke Javanese, they seemed very much at home. 'It's good here,' one man in his fifties told me. 'Once it was wild. Now it's populated.' And, indeed, it was. Young children ran around by watercourses chasing one another and rushed excitedly to the village center when the local ice-cream seller turned up on a moped with a freezer box. It was a glimpse of happiness in a society that was growing both economically and physically, but at the expense of the rainforests and their original inhabitants. The scene revealed just how swiftly dramatic changes could be normalized. Only a couple of decades previously this now densely settled landscape was wild forest with elephants and tigers.

The pressures for land to grow food and produce commodities for export, and the logging that often preceded it, has led to the near complete loss of primary forests from the southern and central lowlands of Sumatra. In Pekanbaru, the state capital of Riau province, I saw the carcasses of burned forest trees next to new houses in the suburbs, revealing just how recently the conversion had occurred there. Yet, for all the deforestation, these rural parts of Sumatra often felt surprisingly empty. Relatively little land was being used for people to grow food for local consumption. Instead, it was being harnessed for the large-scale production of export commodities—mostly palm oil and pulpwood.

Thailand and Cambodia

As a result of logging, resettlement and the conversion of forested lands for plantations and small-scale farming, much of Asia's rainforests have already gone. This is not just the case in Borneo and Sumatra but also most of the rest, in Indonesia, the Philippines and the Malay Peninsula.

The position is equally urgent in Indochina, especially Thailand and Vietnam, where the forests have been heavily logged for the export of wooden furniture, and large areas of land subsequently converted to coffee growing. Such has been the speed of the rise of coffee here that in less than ten years Vietnam has come from nowhere to number two in global coffee production. And, as its forests have been depleted, so sawmills have moved across the borders into neighboring Laos and Cambodia.

As in Central America and Africa, civil conflict has also played a part in the recent deforestation. During the early 1990s the London-based group Global Witness published an investigation into how revenues derived from timber exports from Cambodia were being used by its government to fund a brutal civil war. In the wake of their report, international pressure led to the 1995 introduction of a ban on timber exports. Further investigations revealed, however, that large quantities of rainforest wood continued to be exported via neighboring Thailand, fueled by widespread corruption at all levels in the Cambodian government, which it was estimated was earning between about $10 million and $20 million per month. Much of the wood used was made into garden chairs and tables for European export.

The construction of dams and other large-scale infrastructure has added to the pressures bearing down on Indochina's remaining rainforests. Between 1973 and 2009 the Mekong river basin lost nearly a third of its forest cover. Vietnam and Thailand saw the biggest scale of forest destruction, each losing about 43 percent during that time. Despite this, pockets of intact rainforest landscapes survive in Cambodia, Myanmar and Laos. They need urgent protection, not least to preserve the region's incredibly rich wildlife. Amazingly, over the last fifteen years more than 1,700 species, including birds and mammals, have been newly described in the region.

Queensland deforestation

A mix of population growth, poverty and corruption has fueled the loss of tropical rainforest in many developing countries in the Asia and Pacific region. So it is instructive to compare the situation with what has happened in one of the richest and most developed rainforest nations: Australia.

Tropical rainforests once covered the northern part of Queensland, along the coast facing the Great Barrier Reef, where a unique flora and fauna evolved. About half of this primary forest is now gone, cleared to make way for sugar cane, bananas and livestock, causing problems for forest wildlife, as well as for the world's largest coral reef, which is affected not only by global warming but also run-off from deforested land. The Queensland deforestation shows very clearly how a country needs not only a strong economy and good governance to keep tropical rainforest intact, but firm political intent to protect these ecosystems.

The history of Queensland is as dark as that of the Americas in the way that it treated the indigenous forest-dependent societies. Before the arrival of Europeans, Australia's tropical rainforest was one of the most densely populated parts of the continent. A group of tribes collectively known as the Jirrbal had lived there for thousands of years, their society and economy interwoven with the forests. As was the case in the Americas, initial contact killed much of the indigenous population through transmission of diseases—often just the common cold. Others starved when they were denied access to traditional lands or were shot by settlers. And, as was the pattern in the Amazon, many of the survivors were forced into slavery or servitude, or placed in Christian missions, where they were compelled to abandon their languages and culture.

New Guinea land rights

Most other hunter-gatherer groups across the Asia-Pacific rainforests, whose cultures and way of life depended entirely on intact healthy forests, fared equally badly. The only significant community to survive is on the island of New Guinea. On the western side of this vast and still largely forested territory is the Indonesian province of West Papua. On the eastern side is the separate country of Papua New Guinea.

Dissected by high, snow-capped mountain ranges and ravines, river valleys connect segments of the interior with the coast, while some highland plateau are effectively cut off from the outside world. In its many landscapes are hundreds of different tribal groups, the largest diversity remaining on Earth. These circumstances also make New Guinea the most linguistically rich place in the world, with around 800 living languages.

Papua New Guinea is one of the few countries where tribal peoples still comprise a majority of the population. And they have legally recognized land rights, which explains in part why there's still a lot of rainforest left there. Logging and oil palms are nonetheless a threat, and so is mining for the rich mineral deposits found there. These threats were rendered more potent in 2010, when the Papua New Guinea Parliament passed an amendment to protect corporations from litigation relating to environmental damage and landowner abuse, with all enforcement action decided by central government. Even with this attempt to roll back the ability of tribal societies to protect their land, however, the precipitous mountainous terrain and relative inaccessibility of many landscapes has made the process of deforestation harder and slower compared with the region's other resource-rich forested countries.

This is just as well, for the rainforest landscapes of New Guinea and neighboring islands support some of the most biologically rich ecosystems on Earth. A 2005 scientific expedition to the

Wabumari Village in southeastern Papua New Guinea lies between the ocean and the rainforest (Cool Earth)

mist-shrouded Foja mountains, located in Indonesian territory on the western side of New Guinea, found in just one month of fieldwork dozens of 'new' species, including twenty previously undescribed frogs and a new species of honeyeater bird. The researchers found the area to be an untouched natural wonder with giant crowned pigeons, cassowaries, tree kangaroos and birds of paradise that had not been seen by scientists for decades. Expedition co-leader, Steve Richards of Conservation International, said that the dripping moss forests of the Foja mountains were 'one of the last places on Earth where humans have failed to make an imprint.' Fortunately those forests are protected, at least on paper.

In Papua New Guinea, too, recent scientific investigations have confirmed the rather patchy state of our biological knowledge of the rainforests. A 2009 expedition to the island of New Britain, staged by Conservation International and the local Institute for Biological Research, found 200 'new' species, including previously unknown mammals and two dozen frogs. Among the discoveries were species assigned to their own new genera, including a kind of high-altitude mouse with a two-tone tail.

16 HOW TO DESTROY A RAINFOREST

Indonesia largely erased its rainforests over two decades, aided by the World Bank and the IMF—and multinationals

Indonesia's rainforests are perhaps the most critically endangered of any in the world—imperiled in proportion to just how fast the country has grown, with its rapid development paid for in large part through the liquidation of the country's natural capital—the export of timber, agricultural products, minerals and fossil fuels. Along the way, the tropical rainforests have almost disappeared across much of the country. I made my first visit to Jakarta in 1993, when the Indonesian capital was embarking on a construction boom, with ranks of new concrete and steel skyscrapers filling the skyline. Returning two decades later, I found traffic-choked highways connecting the smart new suburbs of a sprawling megacity, with high-rise homes and offices revealing dazzling economic expansion.

There were still traces of old Jakarta in the fine seventeenth-century buildings that were once at the heart of Batavia—the old colonial capital of the Dutch East Indies. Beneath the new flyovers were mini shanty towns, providing reminders of Indonesia's more recent history. They housed some of the legions of poor who inhabited not only many of Indonesia's rural areas but also its largest cities. The informal settlements revealed how the benefits of that rapid economic growth had been unevenly shared. And the signs were everywhere. On the huge multi-lane highway above the rough dwellings a young man drove a brand-new yellow Porsche, his sleek machine passing a wiry old man cycling along the side, carrying a bundle of sticks on his back.

I saw how the impacts of economic development on the forests were still continuing. Flying on to Palembang, the capital of South Sumatra, the plane passed over oil palm and pulpwood plantations that spread to the distant horizon. Drainage ditches and roads criss-crossed the landscape, taking no heed of natural features. The only landscape elements that had resisted were the meanders of major rivers that threaded between the endless ranks of palms, carrying thick loads of reddish-brown eroded soil toward the sea.

On arrival in Pelambang I walked by the banks of the Musi river and the art deco steel bridge built by Dutch engineers during the 1930s. Thin black strips of black ash slowly drifted down through the yellowish hazy sky. It was like some kind of macabre confetti, marking the union between rampant economic development and environmental destruction. Those bedfellows had given birth to a good life for some, but at growing ecological cost. Through the steady drop of fallout from forest and peatland fires burning across the countryside around the city, a huge black barge headed downstream toward the coast. It was loaded with coal that would be piled into a power station generating the electricity needed to maintain the country's continuing economic growth, illuminating Jakarta, other cities and the country's burgeoning middle classes.

Ashes to ashes

I was in Sumatra to make a film about rainforest conservation, although I feared I might be several years too late. Heading north from the city I saw the enormous scale of peat swamp

Rainforest cleared to make way for oil palm plantations. West Kalimantan, Indonesia (Greenpeace)

drainage that had taken place to make way for farming and forestry. Trucks piled high with fruiting heads of oil palms bumped along the potholed road amid clouds of dust. Next to a palm plantation stood ranks of storage tanks, the size of oil depots you see at a port. Down a dusty track a pall of smoke rose from a mill processing the palm heads into oil.

There were great blocks of rubber too, piled by the side of the road and looking like lumps of dirty snow. Barges impossibly overloaded with logs from pulpwood plantations headed down the Musi river to a pulp mill. And above it all was a haze from the burning forests and peatlands—a deadly yellowish veil that caused many to wrap scarves around their faces.

It was August—the dry season—and a time when there was always a risk of wild fires. That year, the risk was compounded by prolonged drought and high temperatures—and the fires got utterly out of control. Our group came across one.

It had been deliberately started in an area of good-quality forest, most likely to make way for a plantation of oil palms. It was thus totally illegal and in just a few hours an area of about 2 square kilometers was up in smoke. The trees were destroyed and the land beneath reduced to ashes. It had been a peat swamp forest and when those ecosystems are set alight the fire gets into the ground and can burn for months, smoldering in the peat to re-emerge when apparently extinguished. We walked into the recently scorched area with caution, having been told that fires on deep peat sometimes burn out caverns below ground that can swallow a whole bulldozer. It felt like walking in soft mud, except that intense heat came through your wellington boots, making you wonder if they'd melt onto your feet.

The fire had broken out next to a pulpwood plantation, who were concerned about their own operation. To make the land usable for pulpwood—an intensive forestry operation—drainage canals had been cut across the landscape on a grid that extended over hundreds of square kilometers. This had, as intended, dried out the land and the plantation managers were doing their best to control the fire, lest millions of dollars worth of damage were caused. A helicopter with a water bucket slung below commuted back and forth, while a firefighting team on the ground used a pump mounted on the back of a pickup truck. Jets of warm brown water were powered forth, hitting the hot ground and bursting into clouds of steam and ash that mingled with the smoke from the peat and trees.

The fire crew was evidently well trained, and well equipped too, with new safety gear, modern vehicles and a drone that flew above the fire sending live pictures to a laptop, helping the team to plan their attack against the blaze. But, even with all the hardware, the best they were able to do was to contain the fire on the side where it threatened the plantation. They had no chance of putting it out and it would undoubtedly lead to the loss of this particular patch of forest. Although it had been logged previously, the forest would have sustained some of the local wildlife, perhaps even including remnant populations of tigers and elephants. No longer.

Despite the best efforts of firefighters, an area of rainforest is lost to fire in a couple of hours. Bare tree roots reveal the loss of about 2 meters of peat (TJ)

The fire was one of thousands tracked that year by satellites monitoring the forests of Sumatra and Borneo, and which led to daily carbon dioxide emissions in excess of that of the entire United States. I looked around the scene of devastation and saw the black skeletal roots of the once massive forest trees that now stood some 2 meters above the new, and much lower, land surface. Land had become atmosphere. Carbon that was a few hours before tied up in peat and trees was now carbon dioxide in the air, as well as microscopic particles and smoke that would threaten the health of people, including the millions of inhabitants on the other side of the Straits of Malacca. Television pictures that year showed a dense toxic haze enveloping the streets of the downtown districts of Singapore and Kuala Lumpur.

On top of health-threatening air pollution these peat forest fires produce a lot of soot—so-called black carbon. This gets high up into the atmosphere and falls out thousands of miles away, including on the snowfields of the Himalayas, where it hastens melting, causing the ever more rapid retreat of the glaciers. And, while tiny particles facilitate cloud and rain formation (as we saw in Chapter 1), smoke doesn't help. On the contrary, the soot and other particles from burning rainforests result in clouds that produce less rain, not more.

While a logged rainforest might be compared to a bombed city, where services and infrastructure are temporarily interrupted, this one had been nuked into oblivion. Reduced to elemental ash and gas, all of its major functions had been destroyed, even the microscopic fungi that once lived in the moist peat soil. A dragonfly and then a butterfly fluttered across the smoking ground, but apart from these transient visitors, who'd now find no means of sustenance, I could see nothing that had endured the intensity of the fire. It would now be planted (illegally) with a sterile monoculture of oil palms and, although eventually turning green again, would support little wildlife.

An abandoned logging camp in Sumatra (TJ)

The last Sumatran tigers

A couple of days after seeing the effects of that peat forest fire, we arrived in the city of Jambi. Night had come and through the gloom motorcycle headlights illuminated shafts of smoke. The air smelled of burning peat and next day, amid landscapes of oil palms, rubber and heavily logged forests, the smoke became ever thicker. The old timber roads that had first opened up the rainforests during the 1970s now gave access not only to plantations but also to open-cast coal mines and small-scale oil extraction using low-tech 'nodding donkeys.'

As we traveled through southern and central Sumatra the process of deforestation seemed largely complete. After driving for days, the first good forest we saw was at the far end of a track hardly passable even by the toughest off-road vehicle. The fact that there was anything left there was down to the difficulty of getting logs out. This and other little areas of forest in a landscape known as Dangtu still had Sumatran tigers, clouded leopards, tapirs and sun bears.

Those forests were clearly very special, but far from safe. In one piece of what had until recently been undisturbed primary forest we saw evidence of small-scale illegal logging. A rough sleeping platform covered with sheets of torn polythene revealed activity, as did the straight lines of cut hardwood planks that had been made up with a portable sawmill, dragged in there by the loggers. The planks were left behind in the abandoned camp along with branches and twigs cut from felled trees. Drying on the forest floor it was kindling for a future forest fire. Why the loggers had departed without taking with them the fruits of their illicit labors wasn't clear. Perhaps the threat of enforcement had moved them on, or it may even have been because of the wildlife. In Jambi I was told a story of a man-eating tiger that had killed illegal loggers. It had come into their camp in the forest and, over several nights, had taken six men, silently with no one noticing until the morning.

Whatever the reason for the departure of the loggers from this camp, the incentives to engage in such activities remain considerable, despite the dangers. Timber from valuable rainforest trees can provide a major boost to income, and in that remote part of Sumatra there were few other options. Villages comprised basic two-room shacks but they had diesel generators providing power, and in one home I saw a young girl sprawled in front of a large flat-screen TV. At the forest edge, near to a few last tigers, she was tuned into images from an entire world, and to adverts aimed at Indonesia's fast-growing consumers. She'd probably want to enjoy the kinds of comforts increasingly accessible to many city dwellers, and her young parents would too. Once that world and its modern wonders are visible, then the risks that come with illegal logging are worth taking.

Logging concessions

It is almost invariably logging—whether permitted and planned by governments, or illegal and undertaken outside any planned activities—that is the first step toward complete deforestation. Government-sanctioned logging is often done through the allocation of concessions, parcels of land in which a specific quantity of trees and species can be taken. In Indonesia these are allocated by the country's forestry ministry, with the national and regional economic interest advanced through the supply of timber, employment by forestry, sawmill and transport companies, and taxes and exchange, as the wood is exported.

Concession agreements can have more or less emphasis on sustaining the forest while taking out logs, but for many logging interests there are financial incentives to engage in free-for-all activities. That might involve taking more trees than are sustainable, with consequences that include the depletion of the timber resource, less wildlife and carbon.

Other concessions are allocated to palm oil companies, pulp and paper plantations, and mining interests, including those extracting coal and iron ore. Once timber has been extracted and the forest degraded to the point where there is not much left to extract, it is burned and palm oil or plantations of fast-growing trees are established for pulpwood. In the process of converting the degraded logged forests to pulpwood plantations, the native species left after the valuable trees have been extracted are fed into pulp mills.

Traveling through Jambi province we passed many recently cleared areas of forest where the charred stumps of rainforest trees stood proud of young plantations of acacias and oil palms. This was planned deforestation, legal and under some level of state control. Less easy to influence was the smaller-scale clearance of forests undertaken through illegal activities. Forest areas that I had seen in South Sumatra, Jambi and Riau provinces during earlier field visits were all subject to pressures from people seeking access to timber and land, and clearance through fire. And behind it all there is often political corruption, when landless and poverty-stricken people are encouraged to occupy land in order to further business investments.

This works as follows. City-based investors pay poor rural people to clear the forests and plant the land with oil palms. The financiers then claim ownership—as previously vacant land is being used. Once property rights are established, the land can be sold for a profit, generating more finance to invest in doing more of the same. Fire is the tool of choice for doing the clearance. The encroachment of unallocated public land is thus a major cause of forest loss. Sometimes the investors ship in families to test the local enforcement capacity. Some non-governmental groups also encourage encroachment, drawing on the rhetoric of the 'right to development.'

The World Bank and IMF destruction of forests

During my travels in Sumatra I recalled campaigns from the 1980s and 1990s that sought to draw attention to the folly of the official transmigration program that moved so many people into these remote areas in the first place. Much of the program had been enacted with external financial assistance, including from the World Bank, which issued loans of more than $500 million dollars to Indonesia. Alongside contributing to social tensions and ecological damage, the World Bank's 'help' became a debt that contributed to financial difficulties.

The World Bank and its sister agency the IMF advocated economic liberalization to facilitate export-led growth as a remedy, not only in Indonesia but in other Asian economies facing similar problems. During the 1990s economic liberalization became an obsession for the IMF, World Bank and OECD, all organizations dominated by wealthy Western nations. In Asia, as in Africa and Latin America, wrenching open economies was a key aspect of adjustment programs aimed at managing otherwise impossible levels of debt. In the case of Indonesia the problem was partly created by the Transmigration program, which cost $7,000 for each family involved.

The new policies were, in their own narrow terms, initially a success. Money flowed in, creating a credit boom for those wishing to borrow against what were regarded as low-risk and high-earning enterprises, including real estate development and natural resource extraction. The rapid expansion of Asian cities— Jakarta, Bangkok, Kuala Lumpur—was one visible outcome. That torrent of money, however, led within a decade to financial crisis. This was in part because the very rapid liberalization of the Asian countries' financial sectors wasn't matched by sufficient regulation as so-called 'crony capitalism' flourished, with loans extended without proper scrutiny. At first economies in the region became overheated. But then they began to slow down. The proportion of non-performing loans rocketed, confidence evaporated and panic followed. Banks withdrew finance almost as quickly as they'd handed it out, and in 1997 the economies of a number of Asian

The clearance of tropical rainforest to make way for a pulpwood plantation, Riau province, Sumatra (Greenpeace)

countries, including the rainforest nations of Indonesia, Malaysia, the Philippines and Thailand, collapsed. There were precipitous falls in stock market values and currencies.

During the early 1990s Indonesian colleagues campaigning for the protection of the forests warned that the liberalization of the economy would lead to a massive expansion of oil palm plantations. They said it was coming because of more finance flowing into that lucrative sector and would cause an increase in deforestation. They were dead right about that, although what they hadn't foreseen at that point were the additional dangers that would come with post-crash financial chaos.

As was the case with the earlier Third World debt crisis of the 1980s, it was the IMF that led efforts to prop up the damaged economies and to get them onto some sort of road to recovery. In late 1997 the organization committed to over $110 billion in short-term loans to the affected countries, including Indonesia. The IMF linked the extension of loans to strict conditions, insisting on the privatization of state-owned businesses, cuts in public spending and higher taxes. To be able to repay the IMF loans, countries also needed to find ways to increase their overall national income through increased economic growth. That inevitably meant looking again to the exploitation of their natural resources and increasing the export of (among other things) timber, rubber and palm oil.

Under pressure from the IMF, the Indonesian government lifted a ten-year ban on the export of unprocessed logs. That ban had helped to create jobs in Indonesia in wood processing industries, but in order to promote more exports it was scrapped. That led to an increase in timber exports, earning more foreign exchange, but also creating unemployment and domestic wood shortages; timber prices rose, triggering more illegal logging. And the IMF could hardly have been unaware of the impact of these policies, having reportedly negotiated the terms of their agreement in an office where smoke from burning forests and land clearance for oil palm expansion was visible through the windows.

Not only did the IMF neglect to add conditions that might have safeguarded Indonesia's remaining tropical rainforests, but their loan conditions included cuts in the budgets available to Indonesian agencies charged with forest protection, at a time when pressures were most intense. So the forests burned, fueled by the continuing expansion of palm oil and culminating in that catastrophic dry season of 2015.

As similar policies were put into place across the rainforest countries, there was an increase in exports of many tropical commodities, and inevitably an impact on prices. Just after the

Asian financial crisis hit, palm oil prices plunged—from $600 per tonne in April 1998 to under $200 per tonne by February 2001. The price fall, predictably, produced a vicious cycle. Smallholders, already living on tiny incomes, needed to increase their earnings and often the only way for them to do so was to increase production through further clearance.

I saw this for myself on a trip through tropical rainforest in southern Thailand in 1999. I was at a place called Khao Nor Chu Chi with conservationist Marcus Kohler to assess how the Gurney's pitta, a critically endangered rainforest bird, was doing. The area of forest was already very limited due to encroachment, and just before we arrived it had become even more so. Walking in silence through deep forest with binoculars tightly gripped, ears tuned to any sign of the secretive bird, we suddenly found ourselves in bright sunshine and looking at a couple of hectares of recently felled trees. Whoever had cut the forest had sought to avoid detection by clearing a patch in the middle, rather than at the edge or by a track. Later that evening we talked with some of the local farmers. They told us how the then recent falls in the price of palm oil and rubber (which had also reached a low point) meant that they needed to have more land upon which to grow more of these crops. And so the last little bit of habitat of one of the world's rarest birds was further depleted.

Multinationals, paper tigers and greasy palms

Between the smallholders trying to make ends meet and countries striving to balance their national accounts were giant multinationals looking to maximize profits. During the 1970s vast swathes of forested territory across Sumatra and Borneo were allocated by the Indonesian government to major commercial interests, including paper companies. While they increasingly came to rely on plantations to feed their mighty pulp mills, the companies initially took wood from lowland tropical rainforests.

The giant Parawang pulp mill, Riau province, Sumatra (TJ)

Much of this had already been logged but, even degraded, it was a valuable wildlife habitat and carbon store (especially when growing on peatlands). During the late 1990s Friends of the Earth began to research the connections between companies clear-cutting Indonesia's natural rainforests for the paper industry and the Western finance bankrolling them. The result was a report we called 'Paper Tiger, Hidden Dragons,' published in May 2001, which aimed to put pressure on the banks and investors behind deforestation in Indonesia's paper sector.

My colleague Ed Matthew led our effort to expose the web of connections that enabled a company called Asia Pulp and

Paper (APP) to raise capital in London and New York to pay for the machinery and infrastructure it needed to clear-cut the rainforests before turning them into paper. The Indonesian pulp and paper sector had seen more than $15 billion invested over the previous decade, taking the country into the major league of global producers, vying with US and Canadian paper makers. APP was the largest concession holder in Indonesia, in any sector, and destined to become the world's largest paper company.

By then I was Friends of the Earth's campaigns director and spoke at a press conference we'd arranged in the City of London, seeking to convince the financial institutions of the need for change. We highlighted the importance of protecting the forests and their people, and also showed how investments were at risk, as APP was in a financial crisis and had been implicated with illegal logging. We urged the financial institutions to stop putting money into the Indonesian paper industry until it could be certain it was both legal and sustainable and to use their influence to ensure that APP ceased taking timber from natural forests and replaced that with sustainably harvested pulpwood from plantations.

The banks and other financial bodies paid little attention and seemed largely content to continue bankrolling the liquidation of the rainforests. The trees were being effectively turned into money; ethics didn't appear to have much sway, and neither did financial hazard. But a more productive line of attack was opened up by Greenpeace, who sought to expose connections between rainforest destruction and Western household brands. In order to identify the use of native rainforest trees, genetic testing of paper samples was undertaken. Greenpeace looked for traces of DNA from ramins, which had been protected from 2001 under the CITES rules (which had also been invoked to conserve mahogany).

Warm-blooded pollinators It's not only upon insects that rainforest plants rely to shift genetic material between blooms. Birds and bats help, too. Geoffroy's tailless bat (*Anoura geoffroyi*), Costa Rica (above); long-billed hermit hummingbird (*Phaethornis longirostris*), Costa Rica (below).

Seed carriers Diverse animals include keel-billed toucans (*Ramphastos sulfuratus*), Costa Rica (top); brown agoutis (*Dasyprocta variegata*), Tambopata-Candamo reserve, Peru (bottom left); and southern cassowary (*Casuarius Casuarius*), Daintree National Park, Australia (bottom right)

Forest engineers Many primates are also important seed dispersers, influencing the structure of the entire forest in the process. Lion-tailed Macaque (*Macaca silenus*), India.

Predators A wide range of predators feed on smaller creatures. A turquoise jay (*Cyanolyca turcosa*) eats a butterfly in an Ecuadorian cloud forest (above), while a praying mantis dines on similar fare in the Cuc Phuong National Park in Vietnam (below).

Cold-blooded killers Reptilian predators are legion. These two, both in Costa Rica, are a plumed basilisk (*Basiliscus plumifrons*), above, and the yellow variety of the eyelash pit viper (*Bothriechis schlegelii*), which ambushes prey from its hiding place among yellow blooms.

Camouflage One strategy to avoid being eaten is to avoid being spotted in the first place. Great potoo (*Nyctibius grandis*), Brazil (above); lichen katydid (*Markia hystrix*), Costa Rica (below).

Big cats Similar animals have evolved to occupy similar roles in rainforest ecosystems, even though they are in different parts of the world. This tiger (*Panthera tigris*) is from Thailand (above); the jaguar (*Panthera onca*) from Belize (below).

Ape world All seven species of great ape live in the tropical rainforests, and all are at risk of extinction. Eastern chimpanzee (*Pan troglodites schweinfurthii*) eating fruits in Gombe Stream National Park, Tanzania (above). Mountain gorilla (*Gorilla beringei beringei*) in Volcanoes National Park, Rwanda (below).

Bad news for Ken and Barbie (Greenpeace)

It turned out that a large number of such trees were going into APP's pulp mills, and that paper and card including fiber from these controlled species were being supplied to companies that were household names to Western consumers. They included packaging and products for Tesco, Walmart, Kentucky Fried Chicken, Rank Xerox and Hewlett-Packard. Luxury writing and jewelery company Mont Blanc was using APP fiber, as was Tchibo (the world's fifth largest coffee roaster) and Cartamundi (the world's leading maker of playing cards). Further investigations revealed that Mattel, Disney and Lego were using rainforest fiber in their toy packaging, while another found connections between deforestation and the food giant Danone. The Collins publishing group was implicated, too.

These and other brands were invited by Greenpeace to consider a simple proposition: either stop using APP's paper now, or we'll run a public campaign to boycott your products. One to feel the heat was Mattel, a huge buyer of packaging from APP. Activists unfurled a large banner across the front of the company's Los Angeles headquarters featuring Mattel products Ken and Barbie with the slogan, 'Barbie: It's over. I don't date girls that are into deforestation.' Mattel quickly ended its association with APP. Indeed, one after the other, companies identified by Greenpeace canceled their orders until 130 global brands had publicly distanced themselves from the paper company and its role in the destruction of the Indonesian rainforests.

In parallel with a focus on the Southeast Asian paper industry, campaigners had also begun to ramp up pressure on the palm oil companies. At Friends of the Earth we'd been active in highlighting the threat posed by the expansion of palm oil plantations across tropical Asia since the mid-1990s and in 2005 we began a new public campaign with a report called *Greasy Palms*, setting out how oil palm expansion was driving massive deforestation. On the back of that report we began to target brands such as Tesco that bought palm oil for their products. We made connections between the oil palm plantations and the imminent extinction of species, notably the orangutan. We demanded that companies looked at their suppliers and found sustainable sources not linked with deforestation, threats to wildlife or adverse impacts on local people. Again we found common cause with Greenpeace, which used its network to target global brands.

We found that the companies we were seeking to influence usually didn't have a clue who their palm oil suppliers were. Unilever had many brands using palm oil, including some of its most famous, such as Dove soap and Flora margarine. So in July 2008, to highlight the issue, campaigners abseiled down the side of its global HQ in London, dressed in orangutan suits. While the action was taking place Greenpeace's John Sauven and I were meeting nearby. His phone rang. It was Unilever's

Head of Sustainability calling to ask what was required to get his Greenpeace colleagues to end the action. 'Easy,' said Sauven. 'Say you'll stop buying palm oil from companies causing deforestation and we'll come down.'

The palm oil connection had become yet another pressure point on multinationals implicated with deforestation and it soon became apparent that even major companies didn't know who produced the oils in their products and found it hard to discover. When Unilever representatives asked if they could visit plantations supplying oil used in their products, many of the owners refused to see them.

These and many other actions began to elevate pressure on multinational companies producing and buying commodities which impacted on the clearance of Asian forests. Less obvious than deforestation, however, was the decline of the rainforest wildlife.

17 TIGERS, PANGOLINS AND HONEYCREEPERS

Hunting, forest fragmentation and the effects of introduced invasive species are all taking a toll on forest wildlife across Asia and the Pacific

As the first light of dawn began to show on the eastern horizon, I could hear the faraway calls of gibbons. I was in a huge Sumatran plantation whose fast-growing, monoculture trees were destined for one of the world's largest pulp mills. The singing cries of the rainforest apes lifted my spirits. They also gave encouragement to the team of Indonesian scientists I was traveling with, who were conducting a survey into what wildlife remained in a fractured landscape in Sumatra's Riau province. Most of the lowland rainforests here had been logged and then cleared to make way for pulpwood, oil palm and rubber plantations, but there was evidently some wildlife out there.

Walking along one of the many drainage canals cut through these peaty lands, we could see a patch of natural forest ahead. It not only looked but sounded different, with a lot of birdsong coming from the semi-natural area. As we got closer the colorful flashes of bee-eaters hunting insects on the wing became visible, as did the darting flight of a falconet (a sparrow-sized bird of prey) and the shining glossy plumage of sunbirds.

To make sure we recorded as much as possible, we walked in silence, my companions taking photographs with long lenses

Where natural forest meets plantation, Sumatra (TJ)

to document the many birds we could see around that little area of semi-natural forest. Then one of them tapped me on the shoulder and pointed up into a tree. At first I couldn't see what he was looking at, but then made out the coils of a huge python resting in the fork. It was waiting for some unfortunate monkey or bird to come close enough to be caught. A moment later he pointed down excitedly and again it took me a moment to see what he was looking at. Then I saw it—a tiger footprint. The animal had been there perhaps minutes earlier, as dawn broke. This really was special, as was the fact that a little group of elephants had been seen nearby, and our torchlight glimpse, the night before, of a sun bear.

Later on that day, crossing an area of plantation, I could see big black birds flying between a couple of patches of logged natural forest. A closer look with binoculars revealed them to be black hornbills. Their great ivory-colored beaks make them look top-heavy, yet they flew with surprising grace and power on deep scooping wingbeats across the monoculture of trees. They were one of nine species of hornbill in Sumatra and were evidently capable of covering long distances over unsuitable habitat.

The little patch of logged-over forest in which we had seen the tiger footprints was only about 500 meters wide and 3 kilometers long, yet it was sustaining amazing wild creatures, many of which are seriously endangered. With a world population of no more than 400, the Sumatran tiger is on the critically endangered list, and the Sumatran elephant is similarly imperiled, its population thought to be about 1,000 (down from about 3,000 in the past fifteen years). Both these creatures were evidently ranging out from the degraded bit of natural forest and into the plantations, moving through the landscape along the canal-sides and quieter forest tracks. But the little areas of remaining forest, even the quite degraded ones, were key to the fact that such animals were present at all. For tigers they offered a source of deer and pigs, while the elephants could find a wide variety of food plants. The hornbills, too, were flying in and out, still discharging their role as seed dispersers.

Snared

As I'd seen in other rainforests, especially in Africa, the wildlife was not only suffering the effects of deforestation and habitat fragmentation, but also hunting. In that same landscape where I heard gibbons and saw tiger footprints, I was shown snares set to catch deer and pigs. Now and again the trappers would hit the jackpot and catch a tiger. As the snare wire tightens around their neck or a foot, the animals become gripped with panic and either strangle themselves to death or suffer the excruciating pain of a near amputation while they await the trappers' bullet. These big forest cats fetch a lot of money, both for their skins and also for their bones, which are in demand for Chinese medicines.

Forest elephants are killed for their tusks to supply the illicit trade in ivory, while rainforest rhinoceros species are, like their savanna-dwelling counterparts in Africa, poached for their horns, which are prized in Chinese medicine and used to make a powder that is reputedly an aphrodisiac—although there is no evidence for any beneficial effect. Hunting has hastened the decline of the critically endangered Sumatran rhinoceros—the smallest and oldest of the species. They travel through the shady forests foraging for twigs, foliage and fruits, leaving networks of scent trails. Today there are thought to be fewer than 100 left as they hover at the brink of extinction. The Javan rhinoceros has an even smaller population of about sixty and remains in a single National Park.

The trade in wildlife species, including those banned from commerce by international law, continues to grow and widen its spread. Even hornbills have recently been subject to a massive increase in hunting pressure. One of the most magnificent birds remaining in those Sumatran rainforests, the helmeted hornbill, has recently suffered a high level of trapping for its 'red ivory,' derived from the birds' bony casque structure. The trade in carved ornaments and other trinkets is driving dramatic decline.

Pangolins—scale-covered anteaters—are another group of rainforest animals under threat from hunters. They are caught on a massive scale and illegally shipped across national borders, especially to China, where their fetuses, scales and blood are used in medicine, the meat is considered a delicacy and stuffed skins are sold as souvenirs. As hunting pressures and efforts to clamp down on the trade have caused the animals to become scarcer, so the prices have gone up, further incentivizing hunting. In the early 1990s a kilo of pangolin scales would fetch about $10 in China's black market. Today it is around $175.

Even when people living in rural areas don't hunt the local wildlife, there can still be problems. In Sumatra tigers sometimes kill people and then are killed. Elephants, too, are often killed after making incursions into farmland.

As larger animals become scarcer and in some cases are eliminated completely, there are complex knock-on effects, which may not become apparent for decades, or even centuries.

Pangolins are the among the wide range of rainforest animals threatened by the illegal wildlife trade (Getty)

Take those dwindling forests in Sumatra, where once there were rhinoceroses, orangutans and elephants (never mind hornbills, gibbons and bats), all in their different ways shaping not only the present but also the future character of the rainforests. As all of them have declined to remnant populations, there are now very few places where these animals still form an intact guild of forest architects. Indeed, in the whole of Sumatra there is only one landscape remaining where these creatures survive together—the Leuser landscape in Aceh province in the north of the island.

Covering more than 26,000 square kilometers, the Leuser landscape ranges from steamy shorelines to cool montane forests and includes huge areas of peat swamp forest. Shaped over time by volcanic eruptions, fluctuating sea levels and repeated species migrations, it is one of the most biologically diverse landscapes on Earth and one of the few remaining largely intact forest landscapes in all of Southeast Asia. Even here, however, despite its protected status, there is palm oil expansion and illegal clearance eating away at its edges.

The extent to which the last forests of north Sumatra hold as yet unknown biological treasures was underlined in November 2017 with the publication of research confirming the existence of a 'new' species of great ape—the Tapanuli orangutan. Found only in a single high elevation forest called Batang Toru, the population of about 800 animals is genetically distinct from other orangutans living in Sumatra and Borneo. Of the three species of orangutan the most recently described was found to be the most ancient, with the other two descended from it. The identification of this creature brings the total number of known great ape species living on Earth to seven (eight including humans). All of the non-human great apes live in tropical rainforests and five are critically endangered and two endangered. Only we humans are abundant, and with our

Borneo orangutans (*Pongo pygmaeus*), Tanjung Puting National Park, Kalimantan, Indonesia. These apes are critically endangered because of the loss of their habitat, including to huge plantations of oil palms (TM)

wrecking of the tropical rainforests we are responsible for the desperate plight of all the others.

Pacific aliens

There is one further man-made threat to rainforests and their wildlife: the import of alien species of plants and animals. An extreme example of what can happen when it goes wrong can be seen far out in the Pacific Ocean where the Hawaiian Islands host the most remote tropical rainforests on Earth. In some respects they are well preserved. Dense dark-green cloaks still smother primeval landscapes where rainbows plunge into deep steep-sided valleys and waterfalls cascade through rugged ravines born from molten forces. Warm humid air rolling from the ocean is lifted aloft by soaring cliffs and sunlit volcanic ridges to create slate-grey skies that produce the torrential rains and near permanent mists that sustain moist forests unique to this isolated archipelago.

These little outposts of green set in the planet's greatest expanse of water were never joined to any continent but spawned from mountains that rise some 9,000 meters above the sea floor. All the native land animals and plants found here today are thus descendants of species that managed, one way or another, to get across thousands of kilometers of open sea to forge colonies in the middle of the vast Pacific Ocean.

Wildlife that made it to the Hawaiian Islands was rewarded with an empty world to colonize. Free from the predators, competitors and diseases to which they had evolved behavioral and physical adaptations to survive in their former homelands, the scene was set for rapid expansion and, through natural selection, rapid change. Islands can be biological hothouses that by virtue of their isolation generate new life forms, as founder populations evolve to occupy vacant niches and in the process give rise to new species. This is the reason why 98 percent of Hawaii's native land birds are found nowhere else on Earth.

Hiking up a steep forest trail on the island of Oahu I went to look for one of them, a species of honeycreeper called the Oahu amakihi. It was not only unique to Hawaii but to Oahu, a little bird with green plumage and a long decurved bill. Honeycreepers are believed to be descended from finches that reached Hawaii about 5 million years ago, probably from Asia. The little birds found lush forests with abundant food, an absence of predators and no competitors, and the founder group evolved into some

The extreme decurved bill of Hawaii's i'iwi is adapted for nectar feeding, making the bird a crucial pollinator of native trees (Getty)

57 species, all very likely derived from that one ancient flock, perhaps blown off course in a storm. As time went by, forests that had never before known the sound of birdsong rang out with ever more varied calls.

The level of diversity that emerged among the honeycreepers is as unique as the birds themselves. Even Charles Darwin's classic inspiration for his ideas about evolution, the 14 species of finches on the Galapagos Islands, come nowhere close to the honeycreepers' incredible diversification. Among these Hawaiian gems could be found an example of every songbird bill shape, and some that have no equivalent elsewhere, such as the extreme long decurved sickle-shaped bill of the scarlet-feathered i'iwi, adapted for taking nectar from the flowers that grow on the native ohi'a trees.

This remarkable diversity of native forest birds is matched by that of plants and invertebrates. For example, Hawaiian lobelias have evolved into more than 100 species, including the trees upon which the honeycreepers came to depend. A single original species of tree snail (one can only imagine how that first arrived) has evolved into more than forty different genera (groups of closely related species). All of this and much other diversity evolved in parallel to weave a unique rainforest tapestry.

There was to be one colonist, however, that instead of adding to Hawaii's natural diversity would decrease it. That new arrival was, of course, man. Polynesians first colonized the islands during the eleventh century, having traveled in sea canoes from Samoa. They were followed centuries later by Europeans, the first of whom, in 1778, were British survey ships on an expedition led by Captain James Cook. The successive waves of people cleared forests to make way for farms and ranches, causing widespread deforestation. Far worse for the native wildlife, however, was the large number of fellow travelers that the people brought with them, in the form of non-native mammals, birds and diseases.

This disruption began early. The original Polynesians took pigs, dogs, chickens and rats to Hawaii. Later colonists released other grazing animals and predators, including deer and cats. Adapted for life in their isolated island rainforests the native birds had no fear of or protection from mammals. Rats and mongooses raided nests; cats lurked in dappled shade and ambushed adult birds; and deer, pigs, goats and cattle roamed in the forests, opening up the dense tangled habitats preferred by the native birds.

In the later, colonial period, Hawaiians wiped out the lowland forests and all their native birds, and then sought to brighten up their gardens with colorful species popular elsewhere in the world as pets. Java sparrows, waxbills and ring-necked parakeets were imported and released. They brought with them diseases to which the native species had no resistance and, like the New

World Indians who fell victim to European diseases, local species were decimated by bird malaria and bird pox.

Both of those diseases are spread by mosquitoes, which like the pathogens were also introduced by people. They most likely came as larvae living in barrels of fresh water taken on board ships to provision crews on the long voyages out to Hawaii. In some cases the introduced species multiply each other's damaging effects. For example the pigs create mud wallows in the forest and these collect standing water in which the mosquitoes then lay their eggs. The result is that most of Hawaii's unique honeycreepers became extinct and the majority of the eighteen that remain are considered at serious risk.

Over the past two decades, a further threat has emerged as great swathes of ohi'a forest trees began to die. The cause was found to be a fungus that had somehow reached Hawaii and which can kill a tree in a matter of weeks. Without the ohi'a forests, there will be no habitat for the remaining birds that rely on these trees to supply their nectar-rich diets, hastening still more their decline. The fortunes of the honeycreepers and the forests they inhabit are also linked by the extent to which the birds are vital for the pollination, the trees that they evolved with. The pollination services they provided have not been replaced by the introduced bird species that now live in the forests.

Wider environmental change is also taking a toll. Until recently the cooler rainforests and cloud forests found on the upper slopes of Hawaii's volcanoes offered safety for the remaining native birds, because they were too cool for the malaria-carrying mosquitoes. Because of global warming, however, the insects are moving to higher altitude, leaving fewer and fewer disease-free refuges. Despite conservation efforts, several of the remaining native birds are slipping inexorably toward the final exit of biological oblivion, in the last tropical rainforests in the richest country on Earth.

Where I walked in the hills of Oahu looking for native birds, the tree disease had not yet reached and the forest was still lush and green. It was, however, very quiet. As the wind whistled through the treetops, bird calls and songs were sparse. My search with binoculars and a long wait by a flowering ohi'a with its inviting fluffy red blooms, revealed no sight of the amakihi, just a few introduced species that had colonized the forests on islands that are now regarded as Earth's extinction capital.

The Western Ghats and other Asian hotspots

As in the rainforests of the Americas and Africa, much of the recent deforestation and forest degradation in Asia and the Pacific has taken place in so-called hotspots. Such places are often centered on islands and mountains and these regions of exceptionally rich and unique biological diversity are also defined by being under high human pressure. They include the islands of Sumatra, Borneo and Java, known collectively as Sundaland. In Sumatra and Borneo the process of deforestation has become very advanced, while in Java it is almost complete, with just a few National Parks remaining. Other hotspots, where exceptional biological diversity is at risk, include the forests of Indochina that once spread from Myanmar to the Malay Peninsula; Sri Lanka's surviving patches of forest; and the islands of eastern Indonesia east of Wallace's Line, running through to the incredible biological treasure of Sulawesi and then the Philippines and the East Melanesian Islands.

A further global Asian rainforest hotspot exists through the moist forests that cling to the mountains along the western side of India. The Western Ghats form a long narrow spine running for about 1,600 kilometers from just north of Mumbai south to

Tamil Nadu, close to the southern tip of the Indian peninsula. These uplands comprise a mixed landscape of rolling hills and higher peaks, where a wide range of natural habitats include peat bogs, grasslands and different kinds of forests. Intercepting the southwestern monsoon winds, the very wettest areas are drenched by an astonishing 8 meters of rain each year. In these very moist areas are extensive stands of tropical rainforests and upland cloud forests.

Heading into the hills from the coast it becomes apparent that the land has been subject to fundamental changes. Higher up in the hills one of these is particularly apparent: tea and coffee plantations, established during centuries of colonial rule. The logging of valuable timber trees, encroachment by small-scale farmers, fuel-wood collection, opening land for grazing and the establishment of reservoirs to supply lowland areas have all caused forest loss and degradation, too, while railways and roads have contributed to the further fragmentation of natural habitats.

Some forest remains, however, and it is remarkably rich in wildlife. The Western Ghats cover less than 6 percent of India but host more than 30 percent of the country's mammal, bird, reptile, amphibian and fish species, many found nowhere else on Earth. Walking in those highland rainforests I saw long-tailed macaque monkeys, Malabar hornbills and Malabar parakeets (the latter among 22 bird species unique to these hills), as well as signs of more widespread but declining animals, for example, elephants and tigers. The Western Ghats forests are home to the largest remaining population of Asian elephants and, with more than 570 animals, the single largest tiger population in the world.

About 9 percent of the Western Ghats is protected, including the Periyar Tiger Reserve, where many of the unique plants and animals of this rainforest can still be found. The most impressive

Great Indian hornbill, Periyar Tiger Reserve, Kerala, India (TJ)

creatures I saw there were great Indian hornbills. I encountered these enormous birds in an area of old forest, where a pair fed on fruits in the top of a tall tree. Overlooking them from an elevated ridge I watched them clamber about on thick branches that bent under their weight as they stretched their long bills out to reach the choicest morsels. When they flew their wings made a whooshing sound rather like swans.

Aside from the amazing wildlife found there, these forests are of vital practical importance to what will soon be the world's most populous country. From these cloudy wet highlands flow dozens of rivers, providing drinking water, irrigation and power for approximately 245 million people. India's fast-growing population and economy are, like those everywhere else, 100 percent reliant on fresh water and these forests are a vital piece in the jigsaw that presently keeps it flowing.

For example, it is in these hills that the Kaveri river has its source. Downstream, it flows toward Bangalore, which is almost entirely dependent upon it for water. Bangalore is one of India's most important economic powerhouses, the country's 'Silicon Valley,' but during recent years water shortages have created many stresses and they are set to worsen with climatic volatility. The connection between that city and the wildlife of the Western Ghats hotspot is clear. If the habitat of tigers, hornbills, elephants and a multitude of other species can be protected, and indeed expanded, then the vital water supply of one of India's most vibrant economic hubs might be protected.

18 POACHERS INTO GAMEKEEPERS?

As campaigners became effective at targeting companies causing deforestation, some began to change their practices

I described earlier the deforestation that has taken place across Sumatra—and campaigns by Friends of the Earth and Greenpeace targeting some of the companies responsible for it, such as Asia Pulp and Paper (APP). We left the story (in Chapter 16) with attempts by Friends of the Earth to cut off the company's finance and the Greenpeace campaign that focused on APP's corporate customers. There were early wins as brands like Mattel, Nestlé and Disney took steps to clean up their supply chains. In reaction, APP hired PR firms to put out countermessages that 'there must be a balance between economic and environmental goals.' But the Greenpeace campaign continued to build momentum and a couple of years later APP decided to change its approach—and, remarkably, to explore how they could work together with environmentalists to conserve and reforest areas in which they worked.

In 2010, I was a co-founder of a sustainability advisory group called Robertsbridge which had the mission of 'changing business for good.' Most of my colleagues came from a similar background of environment campaigning and we aimed to use our experience in working with companies to effect positive environmental change. It was, nonetheless, a surprise when APP approached our chairman, Brendan May, to ask if we would assist in implementing their new 'Forest Conservation Policy.'

We were naturally on our guard. APP had a pariah status among the campaign groups and we didn't want to be co-opted into a cosmetic PR initiative. Then again, if they were serious, might we be able to work with APP to establish some new industry leadership, which other companies clearing natural tropical rainforests would have to follow? We talked to Greenpeace about the possibilities. They, too, were cautious but reckoned that APP was genuinely looking to meet their key demands: to stop cutting natural forest and to adopt sustainable forestry practices in their plantations.

New forest landscapes

It was in this odd, unfamiliar context that in May 2013 I climbed into a helicopter to fly above the deforested landscapes of Riau province in central Sumatra. Alongside me were two Robertsbridge colleagues—Charles Secrett (who had first established Friends of the Earth's rainforest campaign) and Brendan May—and Aida Greenbury, a young Indonesian executive in charge of sustainability for APP.

It was a depressing landscape below us—one that revealed precious little evidence of rainforest beyond a fringe of mangroves alongside the brown silt-laden sea. In the place of the natural ecosystems were endless ranks of fast-growing trees destined for pulp mills, to be made into packaging, paper and tissues.

An hour or so into the journey, the pilot announced that we were flying above an area known as Tesso Nilo. I'd first heard about this place in the 1990s as a last bastion of rainforest, threatened by a logging concession. Many timber trees had been removed but it wasn't completely cleared and, known to be a remarkable oasis of wildlife diversity, had been acquired by conservationists. I was shocked to see, however, that the attempt to hold the line there had failed. Lying across the land, the huge stems of trees were scattered in all directions. People had come in, cut the trees, set fire to the felled logs and then established a plantation. From 700 feet up, as the helicopter banked to afford a better view, it was clear that the whole of this huge newly cleared area was planted with young oil palms. This was no amateur effort and, despite being illegal, the encroachment had clearly been backed by considerable logistical capability and no one had intervened to stop it.

I looked toward the distant edge of the 'reserve,' where towering palls of smoke rose up, marking places where the conversion of forest to plantation was an ongoing process. Trees and branches had been bulldozed into huge piles and set alight. Several stretches of the horizon had become invisible as a yellowish haze leaked from the flames. Every now and again a whiff of peaty smoke filled the helicopter cabin. If this place is meant to be a nature reserve, I thought, then god help us.

We sat in silence as the pilot turned the aircraft towards an area where he said we'd find some original primary forests still standing. An hour further on, after another seemingly endless mosaic of swampy second growth, oil palm plantations and extensive areas cleared for pulpwood plantations, we came upon them. Greenbury explained, over the roaring engine, that three months previously APP had called a halt to operations in this and other parts of Sumatra, and across the sea in Indonesian Borneo too. One day they were eating into the remaining areas of natural forest and the next their machinery was silenced.

We flew along one such frontier, where natural stands of trees were left, the fluffy canopy starting where the rows of fast-growing plantation trees and recent clearance abruptly

A protected natural forest illegally cleared to make way for a plantation of oil palm, Tesso Nilo, Riau province, Sumatra (TJ)

stopped. Drainage ditches cut into the peat ended at the natural forest edge, too. There were tigers still living down there, Greenbury said, and if the forest remained so would they—assuming the activities of poachers could be contained.

By the time the bulldozers were stopped, APP had cleared nearly 20,000 square kilometers of natural forest. And that was just one company. On top of this were clearances by a host of other companies, cutting trees for timber, growing oil palm, rubber and trees for pulp, fueled by corruption at every level of government. International conservation groups struggled here, even when they had, in theory at least, legal title to land, like at Tesso Nilo. Forest clearance had been a free-for-all for decades.

This made the APP promise to cease clearing natural forests all the more remarkable. Greenbury pointed to the horizon and told us about a project her company was involved with to protect a last area of natural peat swamp forest. She said that with its buffer zone it occupied more than 7,000 square kilometers. As we headed there the helicopter thudded across a 5-kilometer-thick band of plantations surrounding the reserve and then, across a river, we went in an instant from an industrial landscape to one of the largest remaining blocks of lowland peatland rainforest in Sumatra. The massive reserve was called Giam Siak Kecil. The fact that it was still there was the result of a joint plan between the state government of Riau, the national government of Indonesia and APP. In recognition of its global importance the place had been declared a biosphere reserve by the UN.

Checkpoints run by APP were visible on the rivers and canals at the edge of the forest. These waterways were potential routes deep into the reserve and if they weren't policed they would be used by illegal loggers and palm oil cultivators. Greenbury told us that in addition to the checkpoints her forestry teams had deliberately planted that thick layer of plantations right around the reserve so as to deter poachers, loggers and planters from bothering to make the trip. And if they did, they'd have to use roads that were controlled by APP teams.

All of us experience moments when our world view is subject to challenge, and even change. Sometimes we can dismiss it, but it's harder to do so when you can actually see that there might be grounds to review long-standing opinions. This was one such moment. In the battle for the forests I'd come to regard local people and conservationists as the good guys and big business the enemy. While such a view had been largely justified, the situation on the ground had in some places become more complex. There in central Sumatra farmers had illegally cleared a nature reserve that conservationists couldn't protect, no matter how determined and well founded their efforts. And, while APP had over many years done immense damage to the forests, it was now one of the few actors that had the means to protect a very sizable chunk of what was left.

I could see that in the campaign for the rainforests new territory had begun to open up. While it was essential to be vigilant in the wake of promises that were easy to make and harder to deliver on, it did seem to be the case that APP had changed its ideas and was planning to behave more positively.

Integrated forestry and farming

After the bulldozers fell silent, APP hired experts to survey the remaining natural forests left within its concession areas, to establish their wildlife and carbon values. At the same time, work was carried out to resolve social conflicts. Some of these had arisen decades before, when the national government allocated land to the company without paying much regard for the claims of people already living there. This social and environmental survey work was in turn linked with the company's aim to promote the conservation of some of the keystone wildlife species still hanging on across the lands it managed, including Sumatran tigers and elephants.

APP's initial pledge to cease clearing natural forest proved to be real, and as time went on further commitments were added. These included two important policies for peatland—first, the blocking of thousands of drainage canals to lift the water level, and second taking thousands of hectares of forestry plantations out of production, in the process re-wetting peat previously drained for forestry operations.

Then at the 2015 Paris Climate Change Summit (more on which later), APP announced plans for an integrated forestry and farming program. This ambitious new initiative was designed to promote social goals and improve livelihoods as a strategy for retaining—and expanding—the last natural forests. In order to do this the company established the Belantara Foundation, whose holistic approach would channel money to on-the-ground programs, not only to conserve what was left but to restore natural forests so as to reconnect remaining fragments. The need for this was highlighted in the surveys that followed adoption of APP's Forest Conservation Policy, which showed that protecting isolated bits of forest would be insufficient to achieve wildlife or carbon reduction goals. An approach was needed to embrace the whole landscape.

A glance at maps of Sumatra revealed why. For decades great areas of land had been divided up between different economic interests along arbitrary straight lines that bore no relation to ecological or social realities. One side of a concession boundary might be a pulpwood plantation; the other, oil palm; another line might separate a logging concession from an opencast coal mine. Scattered across these different concessions were patches of natural forest in various stages of degradation and the lines drawn years ago by economic planners paid no heed to the real units of the landscape, such as the watersheds of river catchments or soil types, nor the relative importance of the different blocks of rainforests the concessions would replace—nor, in general, tribal or community boundaries.

Making matters considerably worse was the fact that many of the maps didn't match up, leading to endless disputes between concession holders, with rights to different activities. Under

Giam Siak Kecil Biosphere Reserve is one of the last substantial lowland peat swamp forests left in central Sumatra (UNESCO)

these circumstances a progressive company might undertake to protect a forest remnant only to find it being cut down by a neighbor. And on top of such economic interests were the aspirations of the local people. Many of them regarded land use decisions, taken by a distant central government, as unjust and in violation of their traditional rights. Or they just wanted or needed to make money by occupying 'vacant' land; some of the fires I'd come across in Sumatra were one consequence of that.

On top of this mix were the different government departments—often at odds on national and state levels—who'd allocated the concession rights, and had their own political agendas and constituencies, plus the aims of local, national and international conservation organizations, and international bodies such as the World Bank and the various British, German and Norwegian development agencies funding programs.

All this explained the focus, announced by APP, to pursue integrated farming and forest plans at the scale of entire landscapes. To this end, APP identified ten target landscapes, across the southern half of Sumatra and in West and East Kalimantan in Borneo. Within these landscapes the company would protect the remaining natural forests under its control and seek to work with others to conserve and restore a further 10,000 square kilometers. Part of the plan was to reconnect blocks of remaining forest, some logged years ago, which although degraded had enough wildlife value (including mosaics that still held tigers and elephants) to form the core of restoration programs.

APP put an initial $10 million a year on the table to get the foundation up and running and began to search for partners to increase the funding. The chances of these plans having real impact on the ground were boosted when the South Sumatran government formed an alliance, issuing its own 'green growth' strategy to improve conditions for its rural poor. Positive signals came, too, from the government of West Kalimantan. Should this commitment be backed by supportive policies from Indonesia's national government, then a lot might be achieved, even at a point where much of the original forest has already been lost. A national policy could include the reclassification of logging concessions as permanent forest estate, preventing their legal conversion to industrial plantations. This would be an effective strategy for helping to conserve Indonesia's fast-dwindling forests and could help with plans for landscape-level restoration. Another valuable initiative would be the creation of a single map setting out once and for all the boundaries between the thousands of different concessions.

Agroforestry livelihoods

Alongside such strategies, the key to successful landscape-level conservation is improving the livelihoods of local people. During my travels in Sumatra I met farmers barely able to make a living on the little plots they had. It was clear that if the forests were to have a secure future, then those farmers and their families would have to be more secure. One way to do that is through supporting them to get more from their land, including higher value crops.

To this end APP began conversations with companies and supply lines that might connect the farmers to global markets. Two potential partners were the Italian coffee company Lavazza and the chocolate giant Ferrero. The plan was to experiment with growing coffee and cocoa in a mix of crops, including food for local consumption, plus high-value timber species. This would extract more value from smallholdings. It was a similar approach to that of Cool Earth with the Asháninka in Peru and the projects I'd seen with cocoa farmers in West Africa. In this case, the effort would be targeted among smallholders living and working around the edges of the Berbak and Sembilang National Parks. Among many other things, these protected areas in Jambi and South Sumatra provinces support populations of tigers, elephants, clouded leopards, tapirs and gibbons.

Dr Sonya Dewi coordinates the World Agroforestry Centre's work in Indonesia. She explained to me that a crucial component in agroforestry is helping smallholders integrate trees into the farmed landscape, to achieve a diverse land use system; it is the opposite of the land use trends of recent decades, with their monocultures of palm oil and rubber. Dewi said that taking this approach, where trees are grown for timber on the same plot as a range of crops, enables people to make more money from their land and to be more economically resilient in the face of climate change: 'If one crop fails, others can still contribute to incomes. Agroforestry can also improve gender equality, because if only timber is grown women don't usually get so involved. If there is a range of crops, then there is generally wider participation.' Agroforestry can reduce pressures on the forests, especially in remote areas where people need to be more self-sufficient: 'If people can get food for themselves, fodder for animals, construction materials, fuel and medicines, all from the same plot, they place less pressure on the forests.'

Dewi also told me about a project in Lampong province in southern Sumatra, where smallholders had been supported in growing coffee and fruits under a canopy of timber trees. Because soils were protected from the heavy tropical rains there was much less soil erosion and the rivers once again became clear. There was an increase in the carbon stocks on the land, soil health improved, humidity and soil moisture went up, increasing yields in the process. Incomes rose, too, as the coffee gained premium prices in markets for sustainably-produced crops.

This highly sustainable, positive form of farming is, of course, essentially the same system used by indigenous rainforest people—including in the Amazon before the Europeans arrived. It produced enough food to sustain large cities and its processes improved rather than destroyed soils. It also locked a lot of carbon in the ground—a significant concern in our own century.

But can our modern, globally connected food system find ways to embrace agroforestry? Many experts believe it is possible.

Dewi cited examples where it already takes place but said that 'participation along the value chain is vital, especially in the downstream direction' so as to increase benefits for smallholders by connecting them to the market.

The APP initiative, and these moves toward agroforestry, are encouraging and could form the basis of a new economy that conserves and restores rainforest. But they remain relatively isolated gains, made against a backdrop of continuing deforestation. Too often, company executives and political leaders can't see the deeper economic case—and that the forests really are worth more alive than dead.

PART FIVE

WORTH MORE ALIVE THAN DEAD

19 GLOBALIZED DEFORESTATION . . .

Deforestation is driven by the global economy and policies to maximize growth through the export of natural resources

The campaigns that Friends of the Earth, Greenpeace and many other environmental activists launched in the 1990s to highlight the role of multinational companies in rainforest destruction were not random assaults, moving from one company to another or from commodity to commodity. They were part of a strategy adopted in response to what had been identified as a fundamental shift in power—with the influence of government diminished in the face of the rising power of financial markets and international corporations. This shift was in large part the result of deliberate policies and the emergence of a global economy with reduced barriers to trade—including the trade in natural resources coming from rainforest countries.

It was a process exacerbated, as we have seen, by policies of the World Bank and IMF in reaction to the Asian financial crisis and to the huge debts that had crippled so many developing countries. The ever more familiar remedy—the so-called 'Washington Consensus'—decreed a shift of public spending from apparently unproductive priorities, including the environment, to growth sectors such as roads and ports. It sought to promote exports of natural resources through liberalization of both trade and foreign investment (so that overseas companies might buy local ones), the privatization of state enterprises, deregulation and security for property rights.

Of course, there was the potential for freer trade to be harnessed as a beneficial force for development, too, and this was discussed at the first Earth Summit, held in Rio de Janeiro in June 1992. I attended that meeting with Friends of the Earth colleagues to lobby for a new international treaty for forest protection. A few countries saw the sense of this—notably France and Germany— but many others saw it as simply a hindrance to economic growth. These included many of the dominant G8 nations, as well as several of the rainforest countries—Brazil, Malaysia and India—who regarded forest protection as a threat to their 'right to development.' Meanwhile, as environmentalists were warning of the ecological costs of unrestricted export-led economic growth, moves were afoot to establish a new body to promote just that. The World Trade Organization, or WTO, was signed into being by 123 nations in April 1994. Allied with the 'Washington Consensus' of economic policy prescriptions, it would become a supercharger fitted to the motor of deforestation.

A globalized world: Seattle

New free trade rules require countries to do away with trade barriers. But trade barriers to some are environmental safeguards for others. National or local governments might wish to conserve forests that others see as logs, or vacant land that might be used for soya, oil palms, pulp timber, rubber or cattle. And countries can too easily enter a 'race to the bottom,' where to compete in markets and attract investment they need to cut their existing environmental standards.

The new World Trade Organization, once up and running in January 1995, began negotiations toward a set of agreements that would lead to a rapidly globalized economy. Initial talks culminated in plans for a major summit to be held in Seattle in November 1999 for literally thousands of trade negotiators. Environmental and other campaigners went in larger numbers still to confront the new economic agenda.

I was among them. It was my ninth year with Friends of the Earth and I had become policy and campaigns director in the UK, and was soon to become vice chair of our international organization. This included groups in more than 70 countries, and colleagues from many of them were attending the Seattle summit to raise awareness of the dangers for the environment from globalized trade policies, and to urge governments not to adopt them.

The evening before the talks opened, the San Francisco–based Rainforest Action Network unfurled a banner from a huge crane. It had two arrows—one pointing to 'Democracy,' the other to 'WTO.' The following day was marked by mass protests and acts of civil disobedience. Groups of activists occupied road junctions and enclosed the conference center, arms joined, to prevent the talks taking place. The police used pepper sprays and tear gas as a prelude to baton charges.

Marching with rainforest campaigners from around the world I was suddenly blasted by a stun grenade. It caused bright little stars to spin in my head. Just when I began to gather my senses I was hit in the face by a cloud of gas. People staggered about, reeling from the police assault, as hundreds of arrests were made. I took shelter in a shop doorway, pouring water into my eyes so as to regain my vision. Eventually the police called in the National Guard to take control of downtown Seattle and the following morning the Washington State Convention Centre was finally opened and talks began.

By then, however, the protesters' messages had been heard across the world's media—and, perhaps as important, they had alerted

Seattle protests, November 1999. Proposed free trade rules were seen as a serious new threat to forests (Rainforest Action Network)

some of the attendees from developing countries that not everyone believed the WTO rhetoric about growth, development and overcoming poverty, and that a radical free trade deal opening up their economies to the depredations of transnational corporations might not be such a good plan. The talks collapsed.

The Seattle protesters were branded by the media as an 'anti-globalization' movement, which was true enough of many, but did not portray the whole picture. The reason Friends of the Earth was there was to point out how the process of agreeing global rules to govern an ever more integrated, interdependent and interconnected world had been captured by a particular economic agenda and commercial vested interests. Parallel rules that should have been adopted to protect the environment or uphold social standards had largely been dismissed.

The dangers were all the more grave, as what was proposed would wrench open the economies of countries where corruption was rife, meaning that even if economic growth did result, the benefits would be captured by a small elite. These were among the reasons why for many dissenters in Seattle it wasn't so much the globalization that was the problem but that it was so one-sided—protecting the interests of international companies and political elites, but not the environment, indigenous peoples or ordinary citizens. The movement around this point of view grew and I found myself playing a role within it. A little more than a year after what became known as the 'Battle of Seattle' I was in another of globalization's battle grounds: Davos.

Davos, 2001

Davos is a beautiful place, set beside a lake amid high mountain peaks in the Swiss Alps. It is a fitting location in which to contemplate people's relationships with the natural world and on a bright winter's morning in January 2001 I arrived there to highlight the risks posed by climate change and plight of the tropical rainforests.

Davos hosts the annual meeting of the World Economic Forum, where each winter a lot of important business gets transacted. The chief executives of international corporations, prime ministers, presidents, leading economists and political commentators are among a global elite of invited guests, who gather to consider strategic challenges and find common cause. For most of them the agenda was about global economic growth, driven forward with deregulation and expanding global trade. Over dinner they could also talk through more specific business, such as oil pipeline routes or the financial arrangements that would enable new airports, dams and highway to be built.

The Davos agenda made sense to the leaders gathered there, especially from the particular economic perspective they shared. If you were the chief executive of a major international company under pressure from shareholders to deliver bigger returns, or a politician looking to get re-elected on the back of short-term economic expansion, then it was all perfectly rational. It was increasingly clear, however, to those of us wary of global free trade, that more of this would only hasten the increasingly apparent environmental degradation unfolding across the world, including the destruction of the tropical rainforests.

As the campaign on economic globalization gathered pace Friends of the Earth International decided to shine a light on Davos. It had become clear to me that the outcomes for the Earth's climate, wild forests, coral reefs and declining wildlife would be determined not only by the strength of evidence-based argument, but by the priorities and interests of those who wielded greatest political and economic power. We needed to get our message across at such gatherings.

We didn't expect to get too many one-to-one meetings with the world leaders gathered in Davos, so we had to think creatively how we could communicate to the World Economic Forum why it needed to embrace ecology and use its massive influence to deal with deforestation and climate change, rather than making things worse.

We weren't the only ones seeking to bring a dissenting voice to Davos—protesters from across Europe were arriving for what the Swiss authorities feared might be a repeat of Seattle. The day after we'd arrived, the police had put Davos into lockdown. At Landquart, the train station at the bottom of the valley, anyone who looked out of place was denied passage on trains to Davos and sent back to Zurich. Roadblocks and checkpoints on the main road were similarly intended to prevent protest.

We'd evaded these restrictions, however, by arriving a couple of days before the Forum started. Our colleagues in Friends of the Earth Switzerland had helped to organize a forum called the 'Public Eye on Davos,' at which campaigners would gather to conduct parallel events, to discuss what we believed should be the real agenda for the world leaders—including action to combat deforestation. We'd pitched up to help organize all of that, and to undertake our own protest action.

The idea was simple enough—we would distribute leaflets and get media attention, so that when Forum participants went back to their hotels they'd see on TV another side to the discussion, and if it worked, so would the rest of the world. To make the leafleting interesting to the media, we decided on an ironic juxtaposition of imagery, and to pose as fat cats, decked out with expensive suits, jewelery, cigars and even body padding.

On the morning of the action, we took a nonchalant walk past the Forum entrance. Snow fell gently in huge flakes as we looked at a water cannon truck and obstacles covered with razor wire were placed across the road. Armored police vehicles and soldiers were concentrated around the entrance, and high overhead roared an F-16 jet. Our plan was to get into our costumes and walk down to the security line and hand out leaflets to delegates, and hopefully get some coverage from the news crews that we had alerted. We invited the media to meet us for the start of our action outside the 'Public Eye on Davos' meeting venue.

By the time we were dressed up and ready, the snow had stopped and the sun had come out. We swept into the street to be greeted by a glorious day and six TV crews. A couple of print photographers snapped away too. We really hammed it up, complete with Rolex watches, huge cigars and spoof badges—mine bearing the name of 'Franc Suisse.' Waving to the crews with expansive world leader gestures, we set off down the hill, with BBC World to the left, CNN to the right, Channel 4 and ZDF running backwards in front of us, and started to hand out our bright orange leaflets, focused on vital but overlooked priorities, including stopping deforestation and combating climate change. As we approached the guard

Davos fat cats—the author (right) with Duncan McLaren moments before their arrest in Davos (Friends of the Earth)

posts I noticed a flurry of activity ahead. It was not what we were expecting, however. The water cannon truck began to reverse out of the way and the wire barriers were pulled back by policemen. 'My god, they think we're for real,' I muttered. 'Keep going.' We strode on—Duncan McLaren, our head of research in London, Beat Jans, my counterpart from Friends of the Earth Switzerland, and myself, with our friend and colleague Kerim Yildiz, an international human rights lawyer, close behind.

We walked straight past the guard posts, razor wire and police, waving and smiling, and on into the Forum, where we continued to leaflet. One went to Yasser Arafat, another to Bill Gates. Then the police realized that something was not right.

'Who let you in?' a policeman demanded. 'Stop!' And within seconds the four of us were surrounded, leaflets were snatched away, and we were marched out at gunpoint and detained. Back up the top of the hill, we noted with satisfaction that the cameras were still rolling. That night our protest was broadcast across the world, taking our message—of the need to embed economics within ecology—to the Forum.

The agenda changes

Recalled today, that Davos protest in 2001 feels almost historic, for within a few years the global agenda had changed to the point where our supposedly radical leaflets would look decidedly mainstream. This was partly down to the work of the campaign groups—pointing to the dangers of one-sided economic globalization and publicly highlighting bad practice among corporations. But there was a fast-growing body of ecological science gaining traction in political circles too. By the mid-2000s, climate science—and the urgency of action to combat global warming—became a concern of governments as well as activists, and more and more evidence confirmed the importance of healthy rainforests for human well-being and economic development. The old frames of reference—liquidating the forests for economic growth—could no longer be seen as rational. The corporations, too, began to shift their agendas, as they started to see the risks to their businesses more clearly, not just from activist campaigns that might tarnish their brands, but because of events in the real world—forest fires and droughts among them.

For the rainforests, it was becoming clear that solutions would have to link economics to environmental policies. If the world woke up to the idea that in combating climate change there was no more effective step than halting deforestation, there was a chance of doing so.

20 ... AND GLOBAL SOLUTIONS

As the agenda started to change, informed by ecological science and economic reality, so the machinery of global government began to catch up

Despite the progress made toward legal and sustainable timber and food supplies, the establishment of more protected areas and the allocation of indigenous lands, the rate of deforestation remained on an upward curve through the last decades of the twentieth century. This was because rich rewards continued to be reaped from forest clearance. Timber, palm oil, paper, beef, soya and the rest made huge profits for companies and governments. The forest's 'natural services'— the storage of carbon, circulation of water and maintenance of biodiversity—were not so easily valued. Indeed, many perceived them as without economic significance at all. Adding to this dire economic imbalance was the relative dearth of funding for forest conservation.

Andrew Mitchell, founding director of the Global Canopy Programme (GCP), explained the facts to me in this way: 'Governments have generally been putting up about $1.1 billion a year to try and stop deforestation, whereas the total global value of exported commodities which are causing most of the deforestation is about $135 billion. It's kind of a Newtonian problem, whereby the forests have nothing pushing out that can compete with that wave of money coming in. When you've got one billion against 135, you simply can't win.'

His conclusion, inevitably, raises the question of where those additional billions to protect the forests—and add some balance to the equation—might come from. And over the past decade there have been at least suggestions of an answer. For it has become ever more clear to global scientific advisors, and the governments that they inform, that slowing down the pace of deforestation, and restoring some of the forests that have been lost, offers one of the cheapest and most effective options for the world to slow down the pace of global warming.

The first major breakthrough came in 2007, in Bali.

The Bali climate conference

In December 2007 the thirteenth UN Climate Change Conference opened in Bali. Representatives of 180 countries descended on the island—government negotiators, analysts, scientists and NGOs—with the principal aim of discussing an agreement to follow the Kyoto Protocol, which had come into effect two years previously. This would take the form of an ambitious Bali Action Plan, which included a new forest program dubbed REDD (Reducing Emissions from Deforestation and Degradation). Bali was a fitting location in which to launch such a plan, with Indonesia's own forest degradation dramatically highlighted by those disastrous dry season rainforest and peatland fires that had released billions of tonnes of carbon into atmosphere.

I was attending with Friends of the Earth International colleagues to campaign for an effective Bali plan. By then, fifteen years had elapsed since many of us had been at the Rio de Janeiro Earth Summit, where the original climate change treaty had been adopted. After a decade and a half of talk, there was a palpable sense of urgency. But, while the science

Exasperation. UN climate chief Yvo De Boer (left) next to UN Secretary General Ban Ki-moon (center) and Indonesian President Susilo Bambang Yudhoyono (right) at the Bali conference, 2007 (Getty)

was clear enough, the politics were another thing. The central idea of REDD was basically to offer 'positive incentives' to tropical countries so they'd be less likely to replace rainforests with plantations or cattle pastures. This would not only reduce carbon dioxide emissions, but also bring benefits for water, wildlife and hopefully forest peoples.

The matter of forests and greenhouse gas emissions and the links between them was finally on the agenda, but even among those committed to action for rainforests it was not a simple issue. Many of my international colleagues from Friends of the Earth were concerned about unintended consequences that could emerge from the proposals to pay countries not to cut down their forests. Some feared that REDD would make it possible that a country with a low level of forest loss, such as the Democratic Republic of Congo, could be tempted to increase deforestation in order to get more money for reducing it later. That was certainly a potential problem.

Others pointed out how the large-scale and rapid influx of money into rainforest countries had in the past caused more problems than it had solved. They cited the example of the Tropical Forestry Action Plan that was launched in the mid-1980s and how behind its cloak of 'sustainable forestry' there had been many examples of new money paying for more industrial logging in natural forests and pressure on forest peoples. On top of this, the fact that the new scheme was focused on carbon raised fears that this particular value might be prioritized over the conservation of wildlife.

There were also concerns about the effects on people living in and around forests. Would government agencies be tempted to evict them in order to create carbon stores that would attract millions of dollars in international finance? If countries were going to receive big money for forest carbon, their governments might be inclined to place financial concerns over the interests of small farmers and indigenous people living in the forest. Given previous experience, that was quite plausible.

Then again, if the UN didn't do something about deforestation quickly, the chances of meeting the newly proposed and very challenging goal of limiting global average temperature increase to below 2°C, compared with the pre-industrial era, would be effectively impossible to reach. (Let alone the 1.5°C that would subsequently be seen as the safer target.) The basic fact of the matter was that for the past 50 years—indeed, the past 500 years—rainforest destruction had been predicated on the idea that they were worth more dead than alive, whether for timber, new crops or mining. Now, finally, it was being accepted in those international talks that the short-term benefits derived from all of that were far less than the global value of keeping them intact. The forests were worth more alive than dead. Having spent years lobbying UN negotiations to get forest conservation properly on the agenda and addressed through a meaningful intergovernmental accord with real money behind it, this was the biggest thing I'd seen. While many friends and colleagues remained skeptical, I was convinced that we should cautiously engage with this new opportunity rather than oppose it.

I was in any event about to change my role in the fight for the forests. By the time of the Bali meeting I'd done seventeen years with Friends of the Earth, and battling away against giant corporations, multilateral agencies and governments had taken a toll. I could see burnout approaching, and if I was going to be of use in the continuing struggle I needed to find different ways to play a part.

Around the time of the Bali meeting I told the board of Friends of the Earth that as of July 2008 that I'd be standing down as director. I had no particular plans as to where I'd go, but a few opportunities began to appear, including an initiative put in place by HRH The Prince of Wales, who invited me to become an advisor to his newly launched Prince's Rainforests Project. I was delighted to accept.

The Prince's Rainforests Project

Unlike my gutsy campaigner teams at Friends of the Earth, the people working in the Prince's Rainforests Project (PRP) were mainly from the worlds of finance and banking and were planning to engage big businesses as well as governments, scientific bodies and campaign groups. It was hoped people from different sectors could unite behind proposals in that post-Bali context and make a real difference. The Prince was passionate about the tropical rainforests and determined to push his team to get something going. An intensive process of research and consultation was put in train alongside a plan to raise some of those missing billions to provide an economic counter-force for the forests.

In December 2008 I traveled to Brazil to find out if something along the lines of ideas proposed in Bali might find favor there. Arriving at Brasília I was struck by how different things looked since my first visit there in 1990. Back then many of the people were visibly poor, whereas in 2008 they were clearly richer. There were new roads thronging with new cars. Brazil's burgeoning middle classes were evidently enjoying the benefits of the country's recent spell of rapid economic growth, built in large part on the exploitation of natural resources.

The lands around the capital looked different, too. When I'd passed through nearly two decades before, the seasonal savanna woodlands were being opened for soya production. Huge fields now stretched from horizon to horizon. And, as new varieties of soya were developed that could tolerate the more humid conditions in the rainforest lands to the north and west of there, so the soya frontier had spread. It had earned billions of dollars in the process, as had the expansion of cattle pastures, sugar-cane fields and mining.

I met with ministers from across the government: from the departments of environment, finance and agriculture, and spoke to representatives of a new body called the Amazon Fund that was to receive money from international sources to pay for forest conservation programs. It was clear Brazilian ministers were well aware of the pressures on the forests and were prepared to do more to slow down deforestation. They shared international concerns over climate change but also the national and more immediate problem of threats to water security arising from forest loss. If it were possible to find ways to secure the global carbon benefit while they gained national advantages, from international money and also improved water security, then they were interested.

The information gathered from this and many other consultations was fed back into the PRP's melting pot of ideas, as we looked at different ways in which money might be raised—for example, raising funds via carbon offsets bought by small levies on flights (the biggest part of many of our carbon footprints) and insurance policies (a sector that would see benefits from climate stability). In the end we concluded that, while these sources of finance could work, they'd take years to get properly up and running. Instead, we decided to focus our recommendations on an emergency interim financial package to be funded by rich countries to provide stopgap resources.

HRH The Prince of Wales launches the high-profile communications phase of his Prince's Rainforests Project, May 2009 (Getty)

But doing the work on such proposals was one thing; getting political backing for them, quite another. It was with that latter task in mind that in April 2009 the Prince of Wales hosted what turned out to be a landmark meeting in the historic Throne Room in St James's Palace in London. The session on rainforests was arranged to make the most of the fact that a meeting of the G20 was being held in London that week. The key world leaders were in town, and alongside them we had gathered the heads of several key rainforest countries as well as nations such as Norway, which had (along with Costa Rica and Papua New Guinea) shown such strong leadership on rainforest conservation in Bali.

Among those gathered that spring morning were British Prime Minister Gordon Brown, Presidents Merkel, Berlusconi and Sarkozy, and Secretary Clinton representing President Obama. The leaders of the biggest rainforest countries were there too, including President Lula of Brazil and President Susilo Bambang Yudhoyono of Indonesia. The leaders filed through to the Palace's Throne Room and took their places at the table. Before each of them lay a printed copy of the PRP's proposals. The Prince of Wales entered, took his seat and began to lead what was the highest-level political meeting ever to have taken place on the future of the tropical rainforests.

The leaders of the most powerful nations on Earth had been briefed ahead and arrived prepared to take a positive view of what was suggested. There were warm exchanges and nods of approval as they broadly welcomed the PRP document and undertook to set up their own intergovernmental process to refine the different ideas so that they could be taken forward to the Copenhagen Climate Change Conference later that year, hopefully on the basis of some level of international consensus. It was crucial to move beyond the standoffs that had dogged the

Rio summit, when many rainforest countries saw attempts to conserve forests as a threat to their national development.

As I watched the exchanges taking place across the table in that ancient palace I imagined Queen Elizabeth receiving Francis Drake in that very room, in the autumn of 1580, when he'd returned to England from his circumnavigation of the Earth. What tales might he have told her about the riches of the tropical rainforests, the stories of their peoples, the animals he'd seen and of course the money to be made, including from the potatoes that he might have brought with him to show the Queen? In the palace that day, the arc of history had traveled toward a new realization, where we might at last behave as if we'd understood what both science and traditional indigenous wisdom told us about the value of conserving the forests.

Copenhagen and beyond

Officials from the countries attending the gathering that day in St James's Palace held several more technical meetings before, at the end of 2009, the world once again came together for the Copenhagen Climate Change Conference. It was two years after the Bali meeting and was scheduled to conclude a comprehensive global agreement to reduce greenhouse gas emissions, including from deforestation.

I arrived at the meeting with my friend and fellow climate campaigner Thom Yorke, from the band Radiohead. Thom and I had collaborated on the successful Friends of the Earth 'Big Ask' campaign that set out to get the world's first national climate change law adopted in the UK the previous year. As well as making the case for the forests, we went to promote that idea as a good way for countries to take forward international goals.

A lot was riding on the Copenhagen meeting and the mood was fraught. The political sensitivities and differences that had

Arriving at the Copenhagen conference with Radiohead's Thom Yorke
(Charles Clover)

for years prevented a breakthrough were exposed and progress
was elusive. At the end of the two weeks of talks panic began to
set in. World leaders had expected to arrive during the last day
to tie a ribbon on a neat deal, but were instead confronted by
entrenched positions that had hardly moved during the talks. In
some ways things had gone backwards. World leaders literally
ran around the conference center, staging last-minute meetings
in an attempt to rescue the summit from total collapse. President
Obama rushed past me up a staircase. Chinese Premier Wen
Jiabao marched past with a team of advisors, stony-faced, in the

opposite direction. British Prime Minister Gordon Brown sat in a meeting room in crisis talks with fellow European Union leaders. The minutes ticked by, then so did the hours. Scraps of paper with draft words were passed back and forth but despite the last-ditch top-level efforts only a weak non-binding statement of intent could be thrashed out.

It wasn't a total disaster, however, for among the few substantive promises that emerged from Copenhagen was a pledge from a group of rich countries to spend some $4.5 billion on tropical forest conservation during the subsequent three years. That pledge of hard cash had come from the process initiated by the countries that had met in St James's Palace and it signaled goodwill to the rainforest countries. Alongside that very meaningful gesture, with its commitment of real money, there had been progress in Copenhagen toward refining the rules by which that finance (and a lot more in future) would be spent.

It was to be another six years, though, before a substantive deal would be concluded, in Paris in 2015. During that time the story as to why we might wish to redouble efforts toward forest conservation became ever stronger. Minds became focused when 2014 emerged as the warmest year ever, soon to be superseded by 2015, which was warmer still. So did the growing perception of the risks at hand. For, while there had long been talk about the implications of rising temperature and drought, it became ever clearer this was a real and present threat (and it's only become more obvious since, with 2016 emerging as the warmest year ever—the third in a row to break all previous records).

And it was not just governments that were becoming concerned. Those investing trillions of dollars on behalf of pension funds, insurance companies and corporations were anxious that the money they had invested in agriculture in developing countries carried a growing sense of risk—for example, from drought. This was very important, because until there was a perceptible peril to financial capital it was really only the risk to reputations that many investors and executives cared about. That was the kind of risk that could be made very real through the kinds of campaigns run by the likes of Friends of the Earth and Greenpeace, but on its own that had not been enough to do what was needed.

Andrew Mitchell noted the growing realization of the implications of deforestation for major businesses: 'If you're Cargill and trying to get soya out of the Amazon and you get a massive drought, you can't use the rivers for transport. You have to go 2,500 kilometers along roads to get your product to market and that increases costs and takes longer. That sort of disruption is the kind of thing they're increasingly worried about.' Such analysis gradually became more familiar in company boardrooms and businesses in a range of agriculture and consumer goods sectors, including global giants like Nestlé and Unilever. They unveiled ambitious zero deforestation targets and set about sourcing their commodity ingredients used in their products from sustainable sources.

A sea change was taking place. Just how big a change became clear in Paris, where the UN gathered for its 21st annual Climate Change Conference.

Paris 2015: a turning point?

The UN Climate Change Conference takes place annually but, post-Copenhagen, meetings had made scant progress. Optimism was high, however, for the session scheduled to take place in Paris in December 2015. The meeting would see the biggest gathering of heads of state in human history. And setting the scene was the Prince of Wales, by then a trusted advocate for tropical rainforests, who had been invited by President

Hollande to make the opening speech. The Prince called for urgent action, noting that governments collectively spent more than a trillion dollars ($1,000 billion) on annual subsidies to energy, agriculture and fisheries. 'Just imagine,' he said, 'what could be done if those vast sums supported sustainable energy, farming and fishing, rather than fossil fuels, deforestation and overexploitation of the seas.'

Nine years after the Bali Climate Change Conference (and twenty-five years after I had joined the Friends of the Earth rainforest team), I felt that there was actually hope that this would be a turning point, the moment when the world finally got serious about efforts to stop the destruction of the rainforests. And after eight years of discussions, proposals and negotiations, most of the risks identified by campaigners in Bali had one way or another been debated to the point where solutions could be put in place, if the political will was there to adopt them.

Unlike in Copenhagen, the initial mood was positive. World leaders had opened the meeting and set the scene, rather than turning up at the end, and a whole new style of agreement was under discussion, in which each country would adopt its own target reflecting its national circumstances (rather than all countries trying to agree all targets). I was proud to see how the UK's Climate Change Act that the Friends of the Earth campaign had initiated was being cited as an example of what could be done when individual countries were willing to show leadership. For the rainforest countries a big part of what they could do was linked with stopping (or drastically slowing down) forest loss, or indeed expanding forest cover. There was an expectation that the historic polluters like Europe and the US would go further and make deeper cuts in emissions more quickly. But for the first time at such a meeting the big developing countries would make pledges too, including the rainforest nations.

In addition to a flexible approach to target setting there were new voices being heard. Banks, insurance companies, consumer businesses, agricultural suppliers, international agencies, states and cities were taking part in hundreds of panel discussions and side events. There was a palpable sense that the world could tackle the climate change problem. French diplomats worked around the clock, seeking to create and maintain cordial and constructive relations, striving to get the jigsaw pieces organized so that as many countries as possible could get as much of what they needed as possible, in a meaningful global deal.

It was a tough task, but on the evening of 12 December 2015, after several all-night negotiating sessions, delegates gathered for the final plenary session. Figures from the climate change world including Vice President Al Gore, former World Bank economist Nick Stern and French Ecology Minister Ségolène Royal smiled, shook hands and sat together. US Secretary of State John Kerry strolled around the huge room to speak with his counterparts from around the world, including China. Spontaneous applause broke out as Laurent Fabius, the French Foreign Minister who'd presided over the talks, entered.

How different it was to the breakdown in Copenhagen six years before. There was laughter and friendly backslapping as representatives of 196 countries, and veterans of a process that had lasted more than two decades, beamed with delight. At 18.27 GMT the green gavel held by Laurent Fabius fell and the Paris Agreement was adopted. People stood up, they cheered, hugged and laughed. Some cried. After a quarter-century of striving for that moment, who wouldn't?

For the tropical rainforests Article 5 was most significant. This clause set out how countries were to implement and support REDD, and activities related to 'the role of conservation, sustainable management of forests and enhancement of forest carbon stocks in developing countries.' For many developing

Christiana Figueres, Ban Ki-moon, Laurent Fabius and President François Hollande raise hands together after adoption of a historic global warming pact at the COP 21 Climate Change Conference (Getty)

nations, from whom land use change and forestry account for most of their greenhouse gas emissions, this was the mechanism for meeting their Paris Agreement commitments. For the richer nations, there was encouragement to develop partnerships with developing countries to address deforestation and restore degraded forest landscapes, as part of their own target commitments—a form of offsetting, but one that had forest interests at its heart. As well as a new treaty, there were further firm financial commitments, as Norway, Germany and the UK pledged to provide another $5 billion from 2015 to 2020 for REDD programs.

The rainforest countries themselves were also full participants in the process. Ten African countries (Democratic Republic of Congo, Ethiopia, Kenya, Liberia, Madagascar, Malawi, Niger, Rwanda, Togo and Uganda) committed to restore 1 million square kilometers of forest by 2030. Brazil committed to 'strengthening policies and measures with a view to achieve, in the Brazilian Amazonia, zero illegal deforestation by 2030.' Peru's then environment minister Manuel Pulgar Vidal declared that 'Forest countries in partnership with other governments, the private sector and civil society are set for an increased international effort to eliminate natural deforestation and forest degradation in a few decades.'

The Paris Agreement, as many were quick to note, was still a long way from being a specific plan to halt deforestation. And it couldn't on its own solve problems on the ground. It was too high level for that. What it did do, though, was to set a new frame of reference and ensure that the discussion would no longer be dominated by the idea that the forests had to be sacrificed for development and economic growth. The Paris narrative would instead be about how the forests must be conserved to sustain development. It might seem like a simple shift of emphasis, but the importance of this new line of thinking being cemented at the international level in a treaty cannot be overstated. Nor could the adoption of national targets.

Rainforest countries published their national plans and, alongside policies to manage emissions from fossil fuels, pledged to take action to conserve and restore forests. For example, in addition to a pledge to stop illegal deforestation Brazil took on the commitment to restore 120,000 square kilometers of forest. Much of Indonesia's national plan was similarly linked with the conservation of the rainforests and the protection of the peatland soils that had formed beneath many of them.

Looking back, there is no doubt that the years between 2007 and 2015 marked a turning point for global efforts to conserve the rainforests. It was during that period that a new consensus emerged, one that finally appreciated the forests' value if kept intact. In its wake, REDD money began to become available for rainforest countries to reverse policies that caused deforestation, and global commodity companies, too, began to protect the forests needed to sustain their businesses.

21 VALUING NATURE AND RAINFOREST

Companies and governments have increasingly come to accept the economic and practical value of intact natural systems

Spending billions of dollars to avoid trillions worth of damage certainly makes economic sense, but what might be the chances of actually doing this? The year after the Paris Agreement was concluded I found encouragement in what, not so long before, seemed like an unlikely setting—a meeting of the World Economic Forum. In June 2016, fifteen years after I had been marched away at gunpoint in Davos, I was invited to the Forum's meeting in Tianjin, China, to share my views with industry leaders, scientists and government advisors about the value of nature, particularly in relation to the tropical rainforests. I was also to work with a team from Carnegie Mellon University from Pittsburgh and assist them with presenting an extraordinary new set of data compiled from satellite images showing the rapid pace of change that was taking place across the world.

I traveled to Tianjin by road from Beijing and on my way those changes were all too obvious, in the highly industrialized landscape bereft of any wild or natural features. New coal-fired power stations dotted the horizon powering the storming production lines that had sustained China's explosive economic growth. New eight-lane highways tore across the land where huge factories making bricks sat between grids of intensively farmed fields. The sky was heavy with pollution and the rivers the color of the pale degraded soils, as erosion from farmland carried off the country's means of growing food to the sea. The Tianjin skyline was visible from far off through dense yellowish smog. A bullet train shot across a bridge heading to Beijing at more than 300 kilometers per hour.

China's uncompromising determination to generate as much wealth as possible, as quickly as possible, so as to cut the country's grinding poverty, had coal-fired power stations at its beating heart. And why not—after all, coal had driven Britain's transformation from an agricultural to a modern industrial economy. Here, however, industrialization had gone ten times faster and, with so many more people involved, China's revolution was an order of magnitude bigger.

Apart from a few swifts cruising around the underside of a flyover (the same species that each year migrates to the Congo Basin rainforests), there was not a living wild creature to be seen. I recalled a remarkable estimate that 10,000 years ago 99.9 percent of the total weight of land vertebrates on Earth was made up of wild animals, whereas today around 96 percent is believed to be comprised of people and their domesticated animals. Aside from those birds, all I could see there were people and the brick concentration camps of industrial farms, full of pigs and chickens being fed on the billions of dollars' worth of soya imported each year from Brazil, from where rainforests once stood. I wondered if that shocking 96 percent calculation might actually be an underestimate.

In the city of Tianjin there was an impression that at least some of the planners might have realized the scale of the error and made an attempt to redress the balance by bringing nature into the concrete conurbation that had, over just three decades, doubled in size to a population of 15 million people. Trees had been planted alongside some of the main roads and attempts made to create naturalistic settings beside canalized waterways. In one place there was a 'rock face' sculpted from concrete. Next to that stood a herd of deer, made of plastic. This crudely articulated memory of wildness was trapped between a motorway and an estate of twenty-story apartment blocks.

The value of nature

In the vast concrete, steel and glass Meijiang Convention Centre a group of about 2,000 leaders from business, science and politics had been convened by the World Economic Forum for a three-day discussion themed around the idea of a 'Fourth Industrial Revolution.' Founded on high technology and the convergence between artificial intelligence, biotechnology, nanotechnology, biochemistry and powerful new digital platforms, it was clear that a new phase of industrialization was indeed emerging. Delegates met uncannily human-like robots and experienced virtual reality simulations of worlds they'd only ever know from inside a computer. The androids of science fiction had become real, while fantasy worlds created in machines blurred the lines between fact and imagination.

By contrast with Davos in 2001, where in the crisp Alpine setting nature was rich and beautiful but not the agenda, here amid the smog and concrete of Tianjin it was. How times had changed. At least some of the leaders were now listening and finally heeding warnings about the economic and humanitarian consequences of continuing ecological degradation. Some, it seemed, had begun to understand that no matter how sophisticated the new technology, and the central role it must play in meeting environmental goals, it couldn't help improve the state of a world losing the essential and irreplaceable natural services that kept it going.

I delivered this message to delegates via a presentation that I called 'The Value of Nature,' setting out how the economic goals of countries, the commercial aspirations of companies and targets to cut poverty, could not be met in the long term without the benefit of healthy ecosystems and stable climate. I talked about what wild species do for the human world and how the rainforests supplied essential services in relation to food, carbon and water.

With the team from Carnegie Mellon University, including scientist Randy Sargent (who'd assisted NASA in mapping the surface of Mars), I presented data gathered from satellites over more than thirty years that had been combined into a massive new data set called Earth Time-Lapse. The images were stitched together in a time sequence and shown on a huge screen that conveyed more powerfully than any report could ever do how in a few decades we'd transformed the face of our planet.

The satellite sequences revealed the explosive growth of cities like Shanghai and Tianjin, the shrinkage of lakes, spread of deserts, expansion of open-cast coal mines and the loss of forests. Land turning from green to brown revealed the decline of irrigated farming as ground water had been depleted. We showed how the punishing drought that hit Syria between 2006 and 2010, believed to the worst in nearly 1,000 years, was visible from space and how across the north of the country farming had collapsed. It was widely believed to be one of the factors that sparked the conflict there, one of many examples illustrating how environmental changes can have destabilizing consequences in the human world.

A timelapse sequence revealed the changing extent of West African forests and how the Tai National Park, since the 1980s, had become an isolated block in a region that had lost nearly all its natural vegetation. Two other National Parks in Ivory Coast

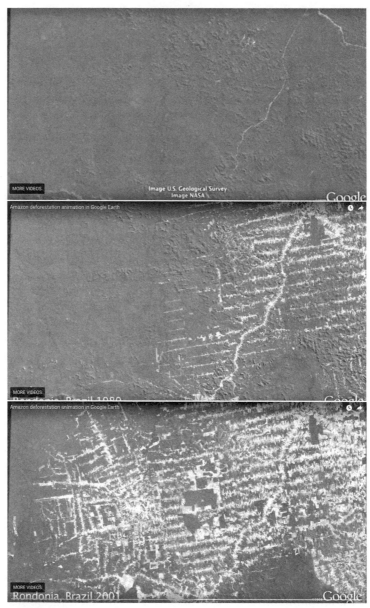

Screen grabs from Google Earth Time-Lapse reveal the effect of a road being built into a remote area of Amazonian rainforest in Rondônia, Brazil, between 1984 and 2001 (Google Earth Engine)

disappeared during the sequence, as civil conflict in that country led to the breakdown of law and they were invaded and cleared by an impoverished land-hungry population.

Another sequence showed the appearance of a new highway into a remote part of Rondônia in the western Amazon basin of Brazil in 1984. Almost immediately in the sequence it was followed by farms spreading along roads that connected to the new long-distance route and into previously untouched primary rainforests, creating that familiar fishbone pattern of forest loss. Zooming out, the time lapse revealed the 'arc of deforestation' spreading across the southern and eastern sides of the Amazon rainforests as roads opened up more and more areas to logging, soya production and cattle pastures. Other pictures from space showed the plumes of smoke coming from the 2015 peat forest fires blazing across Sumatra and Borneo.

I explained the interconnected web of forest issues that first affect the fate of birds, fish and frogs, but gradually embrace the future of a country's economy. I told the story of what I'd seen in Ivory Coast and the new Chinese-built hydroelectric dam under construction there, and how the potential decline in rainfall threatened that project's long-term viability. The highly engineered turbines for this new piece of infrastructure were being built by Alstom at its hydrotechnology site right there in Tianjin. Whether the investors who'd put money into the dam (including $500 million from China's government-backed overseas investment bank) would in the end make any money from that project was down to rain. So it was that the rainforests were directly linked to the industrial explosion that was taking place in the city with the plastic deer, where the World Economic Forum had its meeting.

Having traveled to many tropical rainforest countries and seen the dramatic growth of massive cities, witnessed the rise of their middle classes, watched the rapid expansion of agricultural output and increase in the export of natural resources, there was reason for pessimism. However, there was in those images some hope. From the vantage point of cameras borne on satellites it was possible to see how combinations of growing awareness, new policies, investment choices and technology were altering outcomes on the ground, as changed minds began to lead to changed priorities. At the end of one thirty-year image sequence, compiled from pictures taken from high over a desert in central China, a massive solar power installation, more than 40 kilometers across, suddenly appeared. In another, a Chinese desert went from brown to green as efforts to restore degraded lands had begun to work. Other pictures revealed how deforestation had been halted at the edge of protected areas—for example, across the Amazon region of Brazil, where huge blocks of forests designated as National Parks and indigenous lands still stood out, intact.

As we looked at the images of Brazil, I wondered if in a decade things might look less encouraging. Political chaos in Brazil has the potential to reverse recent progress in slowing down forest loss. Following the impeachment of one president, and with another seeking to cling to political office, and a backdrop of economic difficulties, the budgets of environmental and indigenous agencies have been slashed, causing the implementation of Brazil's forest protection policies to be severely weakened. The increased level of deforestation has already led Norway, the biggest external funder of forest protection in the country, to drastically reduce its contribution.

Then again, in some other countries positive changes were being signaled that might result in less pressure on forests. In the same month that I attended the World Economic Forum meeting in Tianjin, the Chinese government issued guidelines aimed at halving meat consumption. Although this was primarily a public health policy, aimed at reducing the massive economic burden that comes with poor diet, the environmental benefits would also be considerable, including potentially for those Brazilian rainforests, if demand for soya diminished.

Notwithstanding these political ups and downs, the images revealed on the huge screens in front of the leaders did show how environmental damage could be stopped and repaired, so long as the will was there. And the gathering in China offered some reasons for optimism. The World Economic Forum had, just a few months before, taken over coordination of the Tropical Forest Alliance, a government and private sector initiative that aimed to get hundreds of major companies to adopt a collective goal of eliminating deforestation from their activities. The group included many businesses—commodity, consumer goods and food companies—that had been the targets of campaign groups. Some of them still were, but at least they'd now joined a process with a clear end point of zero deforestation. In addition to public pressure, the companies had adopted their new ambitions following interactions with The Prince's Rainforests Project, during the course of which they'd not only come to realize how such a goal was achievable, but also in their interests, for example in relation to water security.

It was another signal that the collective ambition of countries and companies seemed to be laying firmer foundations for action to slow forest loss, at least on paper. The big question was whether this new consensus, and the flow of money and cooperation that came with it, could make a difference where it mattered—on the ground.

22 FUTURE FORESTS

New international agreements, extra money, company policies, public awareness and new data-based tools have transformed the prospects for the tropical rainforests . . . but time is short

The continuing clearance and degradation of the remaining tropical rainforests is not inevitable: it can be stopped and reversed. Right around the tropics there are examples of how things can be done differently. Take the Amani Forest in the Usambara mountains of Tanzania, where 'islands' of vegetation that have remained warm and wet for many thousands of years have been likened to the African equivalent of the Galápagos Islands because of the exceptionally rich wildlife they support.

Professor Neil Burgess, who has spent much of his career documenting life in these forests, told me why the Amani was so important, as the home of 'Africa's only species of tailorbird, endemic frogs, chameleons and many other plants.' He recalled how back in the 1970s some leading conservationists believed it to be doomed: 'It was being logged by a Finnish government-supported operation and also encroached by local farmers seeking out land upon which to grow food and make a living.' The expansion of tea plantations in the highlands posed an additional long-term pressure.

Despite well-founded fears, however, the forest is still there. In 1997 the forest was protected in the 10-square-kilometer Amani Nature Reserve. Despite the pressures imposed on Tanzania by the economic constraints of structural adjustment programs, and despite international finance coming in behind the timber industry, times changed. Today the forest hosts a thriving research station and plans are being laid to reconnect it with other surviving fragments as part of a landscape-level conservation strategy.

The Amani Forest's remarkable turnaround shows how better outcomes are likely to flow from multiple stakeholders being involved—in its case, local communities, the national government, scientific bodies, conservation groups and the international aid community, who switched financial support away from logging and toward conservation. It is evident that multinational companies can increasingly be part of the solution too.

Zero deforestation companies

Take the number of multinational companies that have now adopted zero deforestation policies. Many familiar household brands are among them and they were often influenced to take this step through consumer concerns about the sources of the palm oil used to manufacture their products, from soaps to toothpaste and food products. They include Johnson and Johnson, Unilever, Colgate, Krispy Kreme and Dunkin' Donuts. Others, such as McDonald's and KFC, have been moved to act because of soya's link with forest loss, again highlighted by active campaigning; McDonald's have also committed to deforestation-free beef, as have British supermarket Marks and Spencer. Another group of companies—including Nestlé, Mondelēz, Hershey's and Mars—has adopted zero deforestation goals because of concerns about the future security of cocoa supplies. And many major companies are now also paying attention to where the paper and card comes from in their packaging.

In some ways, these commitments from major companies signal what might over time lead to a process of 'decommodification'—that is to say, an end to the anonymous purchase of ingredients from globalized markets, instead sourcing on the basis of where and how they are produced, attaching standards and achieving positive impact by uniting producers and consumers on the basis of their shared interest in reversing forest loss. The fact that there are now examples of companies finding practical ways to do this confirms that huge possibilities are at hand.

As we saw in Indonesia and Ivory Coast, major companies' zero deforestation policies are increasingly linked with parallel efforts by governments to curb forest loss. Many of the biggest donors of international development finance, including Norway, Germany, the US, France, the UK, Australia and Japan, have, alongside the World Bank and various United Nations agencies, pledged—and are now spending—those billions of dollars to halt deforestation. Across the tropics more than sixty countries are now implementing programs in partnership with the United Nations, funding countries and agencies under the auspices of the REDD. Although still not at anything like the level of finance going into the expansion and intensification of farming, the flow of money has increased.

All the major tropical rainforest countries have REDD programs, and partnerships are now building between governments, companies, local communities and conservation and development groups. Central to the new agenda being advanced by REDD, and now being pioneered by companies, conservation groups and national and regional governments, is action at the level of landscapes.

Nowhere to hide

In parallel with these initiatives has been the development of new technologies to enable greater transparency and accountability, including data gathering and analysis of forest loss and cover. These are increasingly able to reveal change in forest cover in near-real-time, a helpful tool to fight corruption by exposing the reality on the ground. Saying one thing and doing another is becoming hard to get away with.

One of the key technology initiatives is Global Forest Watch, led by the Washington-based World Resources Institute. Crystal Davis, the director of Global Forest Watch, told me about the scope and power of this rapidly evolving tool. It was first launched in 1998, but in the past ten years its capability has been transformed by the US government's bold move to open up the entire NASA Landsat archive. Europe followed suit, providing data from its powerful new Sentinel satellite program and Davis's team has been able to increase its capability still further,

Global Forest Watch provides near-real-time monitoring of forest cover, leaving nowhere to hide for those who might claim one thing and do another (Global Forest Watch)

by working with Google's computing cloud. As Davis explained, 'If you had tried to run our algorithm on a single computer it would have taken about 150 years to analyze the data, but using Google's cloud they were able to do it in four days.'

Such a vast computing capacity is required because the database is at 30-meter resolution (30m per pixel), giving a fine-grained picture of what is going on across all of the world's forests. A look at the main Global Forest Watch map (go to the website and take a look) reveals where deforestation has been most intense since 2000, including many of the places described in the preceding pages, in West Africa, the islands of Indonesia and the southern side of the Amazon basin rainforests.

The author with Jaime Peña, the Asháninka researcher using high-tech equipment to document the diversity of the forest. Jaime explains how he uses a GPS handset to gain precise locations for the images captured by camera traps (Cool Earth)

The power of this detailed analysis of tree cover is transforming the ability of advocates to make the case for forest conservation. 'We understand where deforestation is happening, why it's happening and who's responsible,' Davis explained. Many organizations can use the information being compiled for conservation purposes, including Interpol in their efforts to crack down on shipments of illegally cut timber. Campaigners can use it too, to put pressure on companies, governments and international agencies.

For example, Global Forest Watch helped campaigners generate headlines around the activities of a business called United Cacao. This company was shown to be clearing around 20 square kilometers of intact primary forest in the Peruvian Amazon. It was questionably legal according to that country's laws and definitely in contravention of the company's own sustainability policies. Greenpeace got involved, petitioning to the London Stock Exchange and putting pressure on the government of Peru to undertake an investigation. Forest clearance was halted.

As more data becomes available online, so more can be done through tools like Global Forest Watch. Conservation specialists are adding information to reveal where the most important areas for wildlife are located, potentially leading to better protection as conservation resources are accurately targeted. New tools borne on drones, airplanes and satellites will soon also be able to present more reliable real-time data on carbon, too. Indigenous groups equipped with GPS handsets are also able to use platforms like this to monitor and protect their lands. Policies adopted to protect forests, such as those of the European Union to ensure legal timber imports, will be bolstered, and of course companies that have zero deforestation goals will be better able to succeed and strongly incentivized to do so, given the huge and increasing reputational risks posed by these new analytical platforms. Having your brand linked with deforestation is not good for business—an increasingly clear fact that is focusing minds in boardrooms. Campaigners know this very well too, and the new tools are important resources for the likes of Greenpeace, Friends of the Earth, WWF and Rainforest Action Network.

The Bonn Challenge: forest restoration

Global Forest Watch and other data-rich tools will also be helpful in monitoring progress toward official targets for the *restoration* of degraded and cleared forests. At the global level such goals are set out under the so-called Bonn Challenge. The agreement of time-bound targets for the restoration of forests was first reached at a high-level meeting in the German city of Bonn in 2011, and subsequently strengthened in the New York Declaration on Forests, adopted by the UN in 2014. Under this agreement, countries, non-governmental groups and companies signed up to goals to protect remaining forests and to increase forest cover. The 2011 accord committed signatories to putting in place measures to increase forest cover by 1.5 million square kilometers by 2020, and the 2014 update to 3.5 million square kilometers by 2030. That would be equivalent to expanding the global forest area by about the size of Mongolia by 2020 and by an area the size of India a decade later.

Although these are ambitious goals, progress is being made. At the global meeting of the International Conservation Union in September 2016 I heard from experts monitoring the Bonn Challenge goals that more than 1 million square kilometers of restoration had already been pledged. Included in that area was the million hectares (10,000 square kilometers) announced at the Paris climate summit by Asia Pulp and Paper (APP), the first company to make such a pledge under the Bonn accord.

Imagine what could happen if other companies with major forest footprints, working in partnerships with communities, governments and conservation groups, followed suit? Some companies are already talking about 'net positive' goals in terms of their environmental programs (that is, to leave things better, rather than less worse, as a result of their activities) and the Bonn Challenge is a positive framework through which they can do that, especially for businesses with a significant reliance on tropical commodities or products.

Importantly, the Bonn Challenge has been established not only as a vehicle for the recovery of forests but first and foremost to promote national development. This is for most tropical countries an essential frame for action. It is a fact of political life that poverty reduction generally trumps environmental priorities, so explicitly pursuing forest conservation and restoration because it is good for economic and social reasons is vital. When political decision-makers can see that forests enable water and food security for rural people, in the process helping to prevent migration from the countryside to overstretched cities, they are more likely to sign up to conservation goals.

Underlying the Bonn Challenge is the forest landscape approach, explicitly seeking to restore ecological integrity while at the same time promoting human well-being through integrated and joined-up decision-making.

Protect and conserve

While looking ahead to the development of new tools and frameworks, the future of many remaining tropical rainforests will also rely on old ones, including what has hitherto been our main conservation tool: protected areas. The establishment of National Parks, ecological reserves, indigenous territories and other designations aimed at the conservation of forests have over the decades made a huge positive difference. However, there is limited information on how they are coping with increasing human pressures, especially in those areas that do not enjoy strict top-level protection. What we do know, though, is that when local people enjoy improved social conditions, so the pressures on the protected forests tend to go down—a finding that once more reinforces the direct, fundamental and profound connections between the state of people and nature.

There is also a link with good governance and the fact that nature does better where there is less corruption. The global community might also wish to continue to invest in the integrity of protected forests for reasons of climate change. One estimate suggests that reduced carbon emissions arising from protected tropical rainforests is worth between $6.2 and $7.4 billion per year. That is about one and a half times the total amount spent protecting such places, and of course is a valuation of just one benefit, excluding water, wildlife and all the rest. All of this works better when there is good governance and less corruption.

Having highlighted the broadly positive results arising from the formal protection of forests, it is important to take account of the disputes and conflicts that have sometimes accompanied their designation, for example, in curtailing local people's rights of access. Thoughtful and inclusive planning can resolve at least some such conflicts. For example, in the Periyar Tiger Reserve in India's Western Ghats, where I was able to see magnificent creatures like great hornbills and grey junglefowl, the local guides hired to take tourists through the park were from local indigenous communities. They also had rights to collect honey and small wood for cooking from the forest. These kinds of attempts to provide benefits and protect the rights of those who would otherwise feel excluded or cheated of traditional access are vitally important to get right.

It's not only for the forests that such thoughtful planning is necessary, but also for threatened wildlife species. Periyar reserve is one of many protected places where tigers hang on, and will be a vital part of the plan to not only stop the decline of these creatures, but reverse it. The fact that for the first time in conservation history the loss of tigers appears to have been halted, with about 3,900 now living in the wild, shows that the decline of wildlife species like this is not inevitable. Indeed, all thirteen countries that still have tigers were recently convened by WWF and pledged that by 2022 they would double the number of wild animals.

For both wildlife and key areas of forest, part of the answer is down to the enforcement of existing rules, including those to prevent poaching and maintain the integrity of protected

The Periyar Tiger Reserve strives to combine the protection of Nature with benefits for indigenous communities

areas. As we saw around the borders of the Tai Forest in Ivory Coast, poorly resourced National Parks departments are often fighting ever more intense battles against a rising tide of human pressures. If anti-poaching laws and lines drawn on maps to delineate protected areas are to be held, then sufficient official capacity is needed to do it. That again raises the question of money and who is going to pay. Various developed country governments and international agencies are presently putting up significant finance, especially that linked with carbon emissions reductions, but much more will be needed going forward, as the money so far pledged is spent. What then? One possible answer to maintaining a long-term flow of significant finance could be via carbon charging in the aviation sector.

Offset

Like it or not, the world will be flying more in the decades ahead—and flights are for many in the developed world the largest part of an individual's (and often a business's) carbon footprint. The aviation sector can do a lot to cut carbon emissions and is keen to do so, not least on grounds of cost reduction, using more efficient aircraft and looking at alternative fuels. All this is helpful, but unfortunately not enough. If aviation is going to contribute to meeting global carbon reduction goals, there is also going to be a need for offsets. An offset is basically a way for a polluting sector (like aviation) to pay for action in another sector, so that a benefit equivalent to the damage being caused is created. It is a fraught and controversial area, but one that will be necessary if aviation continues to expand as predicted. A technological breakthrough (such as electric planes, which are under development for shorter routes) would be very welcome indeed, but seems unlikely to deliver in the timescale in which we must take action to reduce carbon emissions.

Some of the money raised from aviation offsetting could be plowed into renewable energy and energy efficiency programs. But finance could also be used to protect a lot of ecosystem carbon. Carbon held in natural systems, including tropical rainforests and the peat beneath them, is vital. Might it be possible to use money from a carbon charge on flight tickets to pay for rainforest protection and restoration? It seems to me that it has to be considered, even if a workable scheme won't be perfect.

In 2016 the United Nations body governing the aviation sector adopted on the back of complex negotiations a new agreement for a 'market based mechanism' to cut emissions. This might soon enable governments and companies to align ambitious goals for emissions reductions from aviation with the REDD program, thereby creating a new funding mechanism. Each year more than 3 billion flight tickets are bought (a number that during the years ahead is expected to rapidly increase) and should each ticket on average have say a $5 carbon charge on it, then that is $15 billion a year. If even a third of that was spent on the conservation and restoration of tropical rainforests, it would make a huge difference to the long-term viability of the REDD projects and programs presently being put into place.

In my capacity as a member of the council of Fauna & Flora International (FFI) I saw how long-term funding will be vital for ongoing progress. In 2016, FFI entered into a project in Liberia, funded by the Norwegian government, to work with smallholder farmers, helping them adopt strategies to increase food production and incomes, using agroforestry methods. It is hoped this will help reduce pressure on some large remaining blocks of the country's tropical rainforests. Zoe Cullen is head of conservation finance at FFI and spends a lot of her time seeking ways to get significant and durable finance behind conservation

Deforested landscape in Liberia. Might carbon charging on flights be
one source of finance to help encourage the protection of the forest
that's left and to restore some of what's gone? (FFI)

work, including for the long-term viability of FFI's activities in
Liberia. A big question in relation to their program there is what
happens in five years' time, when the Norwegian government
money runs out? Will the farmers need to go back to clearing
the forests to make a living, or might there be some other stream
of finance to reward them for forest-friendly behavior? Carbon
money coming from aviation could help to fill the gap.

Cullen is cautious about offsetting, but believes that 'unless
you're actually able to monetize the avoided emissions that

are coming out of a project there's not really much immediate practical benefit for the local people. So we have to have a market for these avoided emissions.' That market (people who want to pay for the benefits being created) could be found under the aviation sector's new UN agreement, linking with REDD as one among a number of longer-term sources of money. Finance would need to be invested against rigorous criteria and standards so as to minimize unintended negative outcomes, but that can increasingly be done, not least using new remote sensing technologies.

While there has been a fierce critique against offsetting aircraft emissions as a means to fight climate change, at this stage it seems to me that we don't really have any choice— not unless we can muster the democratic will to drastically curb flying, or achieve zero carbon aviation technologies. And on the subject of flying and rainforests it is undoubtedly the case (as we saw in Costa Rica) that one way to keep those ecosystems intact is via the economic benefits that can come from ecotourism. If that is to be based on people in different countries going to others far away to spend money to see and experience forests and forest wildlife, then most of them will need to fly. It's a further reminder of how solutions to the various ecological challenges we face need to be integrated with one another. In this case the opportunity is to reduce the net emissions from flights by locking carbon into conserved and regrowing rainforests, which in turn will have an improved economic case for protection because of their potential to attract tourists.

In integrating different approaches I would not, however, suggest that power stations burning fossil fuels should be able to pay to offset their emissions via rainforest protection. Those emissions sources can be eliminated by replacing them with wind, solar, hydro, geothermal and sustainable biomass power. Transport other than aviation also has workable clean technologies available now, including fully electric vehicles recharged with renewable power.

In looking for new opportunities to raise more long-term finance to fund mechanisms for the recovery and conservation of tropical rainforests, a global offsetting scheme in aviation is but one source. Other money could come from investments in sustainable supply chains from multinational companies, including those behind global chocolate and coffee brands, who could do much more to embed agroforestry practices among the smallholders producing their crops. Trust funds, for example, built up from royalties paid to governments by mining companies exploiting valuable ore bodies in or near to rainforests, could help. Small levies on ecotourism would be another source. Lottery money could make a difference, as could fines charged through crackdowns on illegal activities. There could be payments for ecosystem services—for example, from electricity suppliers who benefit from the river flow that drives hydropower turbines, or water companies receiving pure water from intact forests. And of course additional money could be invested by the governments of the wealthier countries, to follow that pledged in Copenhagen, Paris and various points in between and since.

These and other sources of finance could add up to far more than the profits involved in destroying the forest. And such sums devoted to forest conservation would be money very well spent, considering the wide range of essential benefits that would flow to humankind as a result. As we have seen, more money could help pay for the more effective protection of National Parks and in empowering indigenous societies to protect their lands, to scale up agroforestry programs, fund new rural health and education programs, improve governance and help to meet forest restoration goals. If done well, the strategy for the conservation and expansion of the forests could help to address the really deep underlying drivers of environmental decline—through reducing the rate of population growth and by fostering sustainable economic development. All that could be encapsulated in grand visions for future forests.

Big ideas and forest connections

Martin von Hildebrand, the man who demonstrated such vision in his advocacy for indigenous territories in Colombia (see Chapter 10), has set out even more ambitious plans for the wider Amazon basin. When I met him in 2016 he explained to me his idea to create a series of protected areas set amid sustainable farming—a 'network of eco-cultural mosaics,' as he put it, 'an ecological and cultural corridor maintaining the connectivity of the Andean, Amazonian and Atlantic ecosystems' running over the north side of the Amazon basin, from the Andes to the Atlantic coast.

This could only work, of course, through a number of countries acting in concert, so as to combine in one grand scheme all of the Colombian Amazon, all of the Amazon rainforests of Ecuador, a large part of northern Peru, a huge area in the Amazonian region of Venezuela and in northern Brazil, and into the Atlantic countries of Surinam, French Guiana and Guyana. In all, eight main blocks of still largely intact forest would be embraced in a single initiative covering about 2 million square kilometers. Together, they would secure a series of hugely valuable benefits, for those countries and the world, addressing climate change, water security, cultural diversity and wildlife.

I found von Hildebrand's vision plausible not least because some parts of that corridor are already protected. For example, the Asháninka reserve that I visited in Peru lies directly adjacent to the Otishi National Park that in turn connects directly to the vast Manu National Park that embraces a huge expanse

Prince Charles and the Colombian President Juan Manuel Santos visit the Chiribiquete National Park – October 2014 (Getty)

of Amazonian rainforests running up through the subtropical forests to the Andes.

In 2011, during a visit to Colombia, I met von Hildebrand and he told me about his proposal to expand the vast Chiribiquete National Park, so as to link with other protected indigenous and natural reserves. Colombia's President Santos reacted positively to the proposal and in 2013 declared an extension to the park covering 16,000 square kilometers, creating a protected expanse of rainforest of 28,000 square kilometers. This act of leadership helped inspire support from the UK, Germany and Norway, who allocated more than $100 million in reward for reducing emissions from deforestation.

Big visions backed by sufficient money, conservation science, new monitoring tools, community involvement and support, business leadership and political will in national- and state-level government will, of course, be all the more effective in the context of some inspired campaigning. This will always be needed, in order to maintain public support for action and keep governments and companies honest and accountable. Encouragement toward good practice and calling out where action is falling short will be vital in sustaining momentum—for example, among those companies that are making promises to deliver zero deforestation products and governments in meeting the different kinds of spending and conservation and restoration targets they've signed up to.

The fact that I can paint this relatively optimistic picture of the road ahead says a great deal about how far the world has come, and in a very short time. When I began my campaign work for the rainforests back in 1990 I wouldn't have thought such progress would be possible. In the early 2000s, and the time of that demonstration in Davos, political and business will was either absent or weak. Today, however, the case for the rainforests has been successfully made and action is being taken, albeit not yet on a sufficiently large scale, or with the consistency of political leadership that is needed.

Having suggested that a corner is being turned, it has to be said the new road ahead is a long one. But going down it as quickly as we can will be vital for protecting not only the future of the tropical rainforests for their own sake, but also that of people. Should we drive hard and fast, we could secure the bargain of the century—a task yielding massive returns from what is in the greater scheme of things a modest investment.

Those returns would be seen in the multi-trillion-dollar dividend of improved climatic stability, in water security, energy and food security, in helping to reduce the social and political chaos and misery arising from mass migration by advancing poverty reduction and development goals. And beyond all of this lies the task of securing the very future of so much of life on Earth—the evolutionary heritage that will help maintain our food supplies, inspire medical innovation and assist with breakthroughs in design and engineering. It would also be a program to maintain much of human cultural diversity.

We know all we need to know to do this. What it requires now is for all of us who are in a position to help to join in with this historic task. This is not only a call to action for company executives and government ministers, but all the rest of us, in looking carefully at who we vote for, what we eat and consume. For my part, as I put the final touches to this book I'm preparing to head back to the front line, to lead WWF-UK's campaigns. Much of our work will be directed at shaping this continuing story, the ending of which is still to be written. After all, there's no reason why we must continue to watch the inexorable decline of the tropical rainforests. We can save most of what is left and put back much of what's gone, if we want to.

ENDNOTES

Sources and further reading

These endnotes are posted at my website—*www.tonyjuniper.com*—for easy links to the web addresses.

INTRODUCTION: RAINFOREST MATTERS

Friends of the Earth was founded in the UK in 1971. Although early in its history the group ran a campaign for the protection of whales, it was mainly focused on domestic UK issues, such as air and water pollution, waste and British wildlife. In 1984 it embarked on its first truly international venture: the campaign for the tropical rainforests. This in turn became one of the catalysts that helped build the organization's international network, which when I departed in 2008 (by when I was Vice Chair of the global network as well as Director of our operation in in England, Wales and Northern Ireland) was comprised of more than 70 national organizations.

The results of the UN FAO's 1990 assessment that revealed the loss of *moist* tropical forests to be running at about 1 percent per year can be found at *www.fao.org/docrep/007/t0830e/T0830E01.htm#sum*

PART ONE: EARTH'S MOST VITAL SYSTEMS

CHAPTER 1: RAINFOREST—A CLUE IN THE NAME

P. W. Richards, *The Tropical Rain Forest* (London: Cambridge University Press, 1952).

German scholar Andreas Franz Wilhelm Schimper (1856–1901) first used the term *tropische Regenwald* in his founding work on ecology, *Planzengeographie auf physiologischer Grundlage*, published in 1898.

This was translated into English in 1903 as *Plantgeography upon a physiological basis.*

Alexander von Humboldt's travels in the New World are wonderfully documented in Andrea Wulf's *The Invention of Nature* (Knopf, 2015).

Why the rainforest is like a green ocean

Paul Preuss's 2012 article 'Salt Seeds Clouds in the Amazon Rainforest' presents a helpful summary on how cloud-seeding material is produced by rainforest. *www.newscenter.lbl.gov/2012/09/10/amazon-aerosols/*

Research into the role of pollen as a cloud seeding material is explored in a 2015 piece by Nicole Moore, 'Pollen and clouds: April flowers bring May showers.' *www.ns.umich.edu/new/releases/22871-pollen-and -clouds-april-flowers-bring-may-showers*

The proportion of water evaporated from vegetation compared with other terrestrial sources can be found in a 2013 paper, 'Terrestrial water fluxes dominated by transpiration,' by S. Jasechko et al. *www.ncbi.nlm.nih.gov/pubmed/23552893*

Antonio Nobre's excellent 2014 paper *The Future Climate of the Amazon* explains why forests might be dubbed 'green oceans.' *www.ccst.inpe.br/wp -content/uploads/2014/11/The_Future_Climate_of_Amazonia_Report.pdf*

How the rainforests pump water around the planet

A. M. Makarieva and V. G. Gorshkov's biotic pump theory is set out in a 2007 paper, 'Biotic pump of atmospheric moisture as driver of the hydrological cycle on land.' *www.hydrol-earth-syst-sci.net/11/1013/2007/*

For the extent to which the biotic pump is driving large-scale atmospheric circulation see A. M. Makarieva et al.'s 2013 paper, 'Where do winds come from? A new theory on how water vapor condensation influences atmospheric pressure and dynamics.' *www.atmos-chem-phys.net/13/1039/2013/*

V. G. Gorshkov et al.'s book *Biotic Regulation of the Environment: Key Issues of Global Change* (Springer Verlag, 2000) explores how biotic systems maintain conditions for life on Earth.

Peter Bunyard published his research into the functioning of the biotic pump mechanism with co-authors in a 2017 paper, 'Condensation and partial pressure change as a major cause of airflow: experimental evidence.' *www.doi.org/10.15446/dyna.v84n202.61253*

Evidence of increased incidence of drought in the Amazon basin can be found in Marengo et al.'s 2011 paper 'The drought of 2010 in the context of historical droughts in the Amazon region.' *www.inpe.br/ noticias/arquivos/pdf/2011GL047436.pdf*

On the scale of degradation across the Amazon and how this relates to the notion of a 'green ocean' and possible future effects on the hydrological cycle, see Antonio Nobre's 2014 paper, 'The Future Climate of the Amazon.' *www.ccst.inpe.br/wp-content/uploads/2014/11/ The_Future_Climate_of_Amazonia_Report.pdf*

Sky rivers and the long-distance movement of water

An early reference to the idea of sky rivers was made in Reginald E. New's 1992 paper 'Tropospheric rivers?—A pilot study.' *www.onlinelibrary.wiley.com/doi/10.1029/92GL02916/full*

D. V. Spraklen et al write in their 2012 paper about 'Observations of increased tropical rainfall preceded by air passage over forests.' *www.nature.com/articles/nature11390#auth-1*

Deborah Lawrence and Karen Vandecar's important 2015 paper on long-distance water movements and their relationships with tropical deforestation is 'Effects of tropical deforestation on climate and agriculture.' *www.nature.com/articles/nclimate2430*

CHAPTER 2: LIVING ATMOSPHERE

The results of research using ice core data and information gleaned from atmospheric measurements are presented by the British Antarctic Survey at *www.bas.ac.uk/data/our-data/publication/ice-cores-and-climate-change/*

For a full set of references relating to different aspects of climate change, see this compendium of technical resources used in the writing of the *Ladybird Book of Climate Change* (2017) by HRH The Prince of Wales, Dr Tony Juniper and Dr Emily Shuckburgh: *www.rmets.org/ladybird-annex/*

The rainforest carbon store

For a comprehensive compendium of sources on the linkages between deforestation and climate change see *Tropical Forests—A Review*, published by The Prince of Wales's International Sustainability Unit (ISU) in 2015. *www.pcfisu.org/wp-content/uploads/2015/04/Princes-Charities-International-Sustainability-Unit-Tropical-Forests-A-Review.pdf*

The scale of carbon stocks in tropical forests is covered in Sassan S. Saatchi et al.'s 2011 paper 'Benchmark Map of Forest Carbon Stocks in Tropical Regions Across Three Continents.' *www.ncbi.nlm.nih.gov/pmc/articles/PMC3116381/*

A sense of the costs of abating carbon emissions via the prevention and reversal of deforestation is explored in Doug Boucher's 2008 paper 'Out of the Woods: A Realistic Role for Tropical Forests in Curbing Global Warming' (Union of Concerned Scientists). The author suggests the cost of cutting emissions via this route might average at under $3 per tonne. *www.ucsusa.org/sites/default/files/legacy/assets/documents/global_warming/UCS-REDD-Boucher-report.pdf*

The results of mapping of forest loss across the tropical regions during the twenty-first century can be found in M.C. Hansen et al.'s 2013 paper, 'High-resolution global maps of 21st-century forest cover change.' *www.ncbi.nlm.nih.gov/pubmed/24233722*

Two degrees: a global warming danger threshold

Thomas Hilker et al explore the extent to which the Amazon might become a source of carbon dioxide because of drying in their 2014 paper 'Vegetation dynamics and rainfall sensitivity of the Amazon.' *www.pnas.org/content/111/45/16041.abstract*

For research on how some forests release more water vapor under drought conditions, see S.R. Saleska et al.'s 2007 paper 'Amazon forests green-up during 2005 drought.' *www.ncbi.nlm.nih.gov/pubmed/17885095*

Feedback: how forests could create more warming

The impact of severe heat on the Amazon, including the release of carbon, is covered in a 2016 paper by Ker Than, 'Stanford scientists find Amazon rainforest responds quickly to extreme climate events.'

*www.news.stanford.edu/2016/04/28/stanford-scientists-find-amazon
-rainforest-responds-quickly-extreme-climate-events/*

Degradation and fragmentation

The role of large trees in rainforest carbon storage is examined by
J.W. Ferry Slik et al in their 2013 paper, 'Large Trees Drive Forest
Above-Ground Biomass Variation in Moist Lowland Forests Across the
Tropics.' *www.epubs.scu.edu.au/esm_pubs/1893/*

The effects of forest fragmentation on carbon storage are considered
by Sandro Pütz et al in their 2014 paper, 'Long-term carbon loss in
fragmented Neotropical forests.'
*www.producao.usp.br/bitstream/handle/BDPI/46423/Long-term%20
carbon%20loss%20in%20fragmented.pdf?sequence=1&isAllowed=y*

Results from high resolution satellite imagery revealing how tropical
forests are split into about 50 million pieces was published by Katharina
Brinck et al in their 2017 paper 'High resolution analysis of tropical
forest fragmentation and its impact on the global carbon cycle.'
www.nature.com/articles/ncomms14855

For more on how forests might help with storm suppression, see
Antonio Nobre's excellent 2014 paper, 'The Future Climate of the
Amazon.' *www.ccst.inpe.br/wp-content/uploads/2014/11/The_Future_
Climate_of_Amazonia_Report.pdf*

Coral catastrophe; how deforestation impacts on Earth's most diverse marine systems

Research findings into how corals can be affected by sediments released
through deforestation can be found in a 2011 paper by Miriam Weber
et al, 'Mechanisms of damage to corals exposed to sedimentation.'
www.pnas.org/content/109/24/E1558.abstract

Joseph Maina et al.'s 2013 paper sets out why they believe 'Human
deforestation outweighs climate change for coral reefs.'
www.sciencedaily.com/releases/2013/06/130605071714.htm

Threats to the Coral Triangle are explored further in a report by
Lauretta Burke et al from the World Resources Institute, 'Reefs at Risk
Revisited in the Coral Triangle.'
www.wri.org/publication/reefs-risk-revisited-coral-triangle

CHAPTER 3: THE ECOLOGICAL WEAVE

The dynamic nutrient knife-edge

For an explanation on why the Bodélé Depression is such a major source of dust see the 2006 paper by R. Washington et al, 'Links between topography, wind, deflation, lakes and dust: The case of the Bodélé Depression.'
www.onlinelibrary.wiley.com/doi/10.1029/2006GL025827/full

Mycorrhizae: the wood-wide web

For an overview of the important role played by soil fungi in plants' ability to take up nutrients, see James Bever et al, 'Arbuscular Mycorrhizal Fungi: More Diverse than Meets the Eye.'
www.academic.oup.com/bioscience/article/51/11/923/227109

On the role of soil fungi in shaping the diversity of the forest, see Kabir Peay et al, 'Strong coupling of plant and fungal community structure across western Amazonian rainforests.'
www.ncbi.nlm.nih.gov/pmc/articles/PMC3749505/

Zdenka Babikova et al explore how fungi might help to ward off pest attacks in 'Underground signals carried through common mycelial networks warn neighboring plants of aphid attack.'
www.onlinelibrary.wiley.com/doi/10.1111/ele.12115/stract;jsessionid=226B0DF55E9058CEF810C9C1A778EE86.f01t01

An ecological tapestry of connections

The role of bats as pollinators and agents of seed dispersal is explored a little in the Bat Conservation Trust article 'Why bats matter.'
www.bats.org.uk/pages/why_bats_matter.html

Dispersal: how and why the seeds move around the forest

For the seed-dispersing role of red-knobbed hornbills, see Margaret Kinnaird's 2008 paper, 'Evidence for Effective Seed Dispersal by the Sulawesi Red-Knobbed Hornbill, Aceros cassidix.' *www.onlinelibrary.wiley.com/doi/10.1111/j.1744-7429.1998.tb00368.x/abstract*

Research by Ahimsa Campos-Arceiz and Steve Blake into elephants and seed dispersal can be found in their 2011 paper 'Megagardeners of the forest—the role of elephants in seed dispersal.' *www.sciencedirect.com/science/article/pii/S1146609X11000154*

More on how some forest mammals disperse mycorrhizal fungi can be found in Paul Reddell et al, 'Dispersal of Spores of Mycorrhizal Fungi in Scats of Native Mammals in Tropical Forests of Northeastern Australia.' *www.onlinelibrary.wiley.com/doi/10.1111/j.1744-7429.1997.tb00023.x/abstract*

Research into the extent to which the loss of large animals can affect the carbon storage in tropical rainforests is presented in Carolina Bello et al.'s 2015 paper 'Defaunation affects carbon storage in tropical forests.' *www.advances.sciencemag.org/content/1/11/e1501105*

CHAPTER 4: EVOLUTIONARY TREASURES

An ecological superhighway

A summary of some of the impacts and consequences of the formation of the Isthmus of Panama can been found in Dr Helena Fortunato's 2008 paper, 'The Central American land bridge: evolution at work.' *www.schriften.uni-kiel.de/Band%2070/Fortunato_70_56-72.pdf*

For an overview of the history of African rainforests, see Jonathan Kingdom's *Island Africa* (Collins, 1990).

Mega-diversity

For a recent estimate as to the number of species on Earth, see this 2011 news piece in *Nature*: 'Number of species on Earth tagged at 8.7 million.' *www.nature.com/news/2011/110823/full/news.2011.498.html*

James Brown's 2014 paper 'Why are there so many species in the tropics?' explores the question of why there is so much biological diversity in the equatorial regions. *www.ncbi.nlm.nih.gov/pmc/articles/PMC4320694/*

Shore to summit: how rainforest changes with altitude and aspect

Observations in this and the subsequent two sections (*Upland rainforest* and *Land of the frogs*) is based on my research travels in Costa Rica during February and March 2016.

On the restricted range of endemic bird species occurring in the highlands of Central America, Juan C. Martínez-Sánchez provides a helpful overview in his 1995 paper 'Avian Distribution In The Highlands Of Central America.'
www.bio-nica.info/biblioteca/Martinez1995AvianBiogeography1.pdf

Innovation and biomimicry

The blue morpho and how it has inspired alternatives to pigment-based paints is elaborated further in Hayley Dunning's article for the Natural History Museum, 'Insect cells could create everlasting paints.'
www.nhm.ac.uk/our-science/science-news/2014/october/insect-cells-could -create-everlasting-paints.html

Tom Vanderbilt's 'How Biomimicry is Inspiring Human Innovation' introduces the concept on the Smithsonian Institution website.
www.smithsonianmag.com/science-nature/how-biomimicry-is-inspiring -human-innovation-17924040/

An explanation of how glasswing butterfly wings have inspired more efficient solar panels can be found in Valerie R. Binetti et al.'s 2008 paper, 'The natural transparency and piezoelectric response of the Greta oto butterfly wing.'
www.meso.materials.drexel.edu/pubs/Greta_oto_paper.pdf

How Japan's bullet trains were made quieter through inspiration from birds is explained in Tom McKeag's 2012 *Greenbiz* article, 'How one engineer's birdwatching made Japan's bullet train better.'
www.greenbiz.com/blog/2012/10/19/how-one-engineers-birdwatching -made-japans-bullet-train-better

How nature's genetic variety sustains us

The estimate that more than 28,000 species of plant are now utilized in medicine comes from Kew Gardens, 'The State of the World's Plants Report 2017.' *www.kew.org/about-our-organization/press-media/press -releases/the-state-of-the-worlds-plants-report-2017*

Hotspots: where wildlife is richest and most threatened

For more on biodiversity hotspots, see Conservation International's 'Hotspots' report. *www.conservation.org/How/Pages/Hotspots.aspx*

PART TWO: THE AMERICAS

CHAPTER 5: THE 'NEW WORLD' RAINFOREST PEOPLES

The quote attributed to Columbus's journal regarding the demeanor of the indigenous people he encountered in Hispaniola can be found in *The Complete Works of Washington Irving* (1835).

Simon Romero's 2012 piece 'Once Hidden by Forest, Carvings in Land Attest to Amazon's Lost World' in the *New York Times* explores the myth that the Amazon rainforests were untouched wilderness before the arrival of Europeans. *www.nytimes.com/2012/01/15/world/americas/land-carvings-attest-to -amazons-lost-world.html?pagewanted=1&_r=1&emc=eta1*

New World agriculture

An excellent summary of the timing of human colonization of the New World can be found in Charles Mann's *1491* (Random House, 2005). This also covers the spread of disease among New World native populations.

Heather Pringle's *New Scientist* article 'How Europeans brought sickness to the New World' covers similar ground. *www.sciencemag.org/ news/2015/06/how-europeans-brought-sickness-new-world*

For a source for early testimony of Amazonian cities, see Fred Pearce's 2015 *New Scientist* article, 'Myth of pristine Amazon rainforest busted as old cities reappear.' *www.newscientist.com/article/dn27945-myth-of -pristine-amazon-rainforest-busted-as-old-cities-reappear/*

The domestication of pineapples is covered in the Bromeliad Society's 'What Are Bromeliads?.' *www.bsi.org/brom_info/what.html*

For more on the domestication of tomatoes, see Barry Estabrook's article for the Smithsonian. *www.smithsonianmag.com/travel/why-wild -tiny-pimp-tomato-so-important-180955911/*

The domestication of corn is covered in Barry Estabrook's 2015 article 'Why is this wild, pea-sized tomato so important?' *www.smithsonianmag.com/travel/why-wild-tiny-pimp-tomato-so -important-180955911/#k8D3bUmjTEkZ5qYi.99*

Robert J. Hijmans and David M. Spooner's 'Geographic distribution of wild potato species' (2001) has background on the wild ancestors of potatoes. *www.amjbot.org/content/88/11/2101.full*

Soils that sustained cities

For the origins of the *terra preta* soils of the Amazon, see Ute Shueb et al.'s 2016 book *Terra Preta: How the World's Most Fertile Soil Can Help Reverse Climate Change and Reduce World Hunger* (Greystone, 2016).

Good morals

On the natives of Hispaniola being ignorant of metal implements, the quote attributed to Columbus was taken from Howard Zinn's *History of the United States* (Longman, 1980). *www.library.uniteddiversity.coop/More_Books_and_Reports/Howard_ Zinn-A_peoples_history_of_the_United_States.pdf*

The treatment of a captured Carib woman by Michele de Cuneo, and the extent to which that incident reflects on the morals of the time, is examined by Stephanie Wood in *Sex and Sexuality in Early America*, edited by Merril Smith (New York University Press, 1998).

John Hemming's three works on the history of the indigenous peoples of the Amazon are *Red Gold* (covering the period roughly 1500– 1760), *Amazon Frontier* (1760–1910) and *Die If You Must* (1910– present).

The rubber boom

John Hemming's books (above) also include comprehensive consideration of Brazil's rubber boom.

Survival International's article 'Why do they hide?' explains that many 'uncontacted' tribes remain isolated due to the brutal effects of the nineteenth-century rubber boom. *www.survivalinternational.org/articles/3104-why-do-they-hide*

CHAPTER 6: FOREST CLEARANCE IN THE AMERICAS

Theodore Roosevelt's quote is taken from Catherine Caufield's groundbreaking and influential book *In the Rainforest* (University of Chicago Press, 1991).

American clearances

The state of Brazil's Atlantic forests is explored in Ribeiro et al, 'The Brazilian Atlantic Forest: How much is left, and how is the remaining forest distributed? Implications for conservation.'
www.sciencedirect.com/science/article/pii/S0006320709000974

The REDD Desk include good summaries of the status of tropical forests. For Guatemala, see *www.theredddesk.org/countries/guatemala.* For Honduras, see *www.rainforests.mongabay.com/20honduras.htm*

The state of forests in Ecuador is explored by Jefferson Mecham in his article 'Causes and consequences of deforestation in Ecuador.'
www.rainforestinfo.org.au/projects/jefferson.htm

For a fuller explanation of historic deforestation (and not only in South and Central America but globally), see Michael Williams's book, *Deforesting the Earth* (University of Chicago Press, 2009).

The Orinoco and the Amazon

More information on the deforestation rate in the countries of the Amazon basin, including Peru, Bolivia and Colombia, can be found in Mongabay's overview.
www.rainforests.mongabay.com/amazon/amazon_countries.html

Big beef in the Amazon: the soya clearances

The high proportion of tropical forest loss between 1980 and 2000 caused by conversion to agricultural land is covered by Gibbs et al in their 2010 paper 'Tropical Forests Were the Primary Sources of New Agricultural Land in the 1980s and 1990s.'
www.ncbi.nlm.nih.gov/pubmed/20807750

More information on the extent to which agriculture is a driver of at least 40 percent of recent deforestation is presented by the UN's Food and Agriculture Organization (FAO) report, '2016: State of the World's Forests: Forests and Agriculture Land Use Challenges and Opportunities.' *www.fao.org/3/a-i5588e.pdf*

The dramatic drop in the deforestation rate in the Brazilian Amazon between 2004 and 2012 is documented in an article by Rhett A. Butler 'Deforestation drops 16% in the Brazilian Amazon' on Mongabay, which also registers the recent reversal of progress toward reducing the pace of forest loss. *www.news.mongabay.com/2017/10/deforestation-drops -16-in-the-brazilian-amazon/*

CHAPTER 7: COUNTING THE COST: GREEN SHIELDS

Deforestation to recession?

The extent to which the 2014 drought was a factor leading to recession in Brazil is explored in Tim Maverick's article 'Drought Hurts Brazil's Economy, Helps Commodities.' *www.wallstreetdaily.com/2015/02/19/ brazil-drought-economy-commodities/*

Jim Robbins reports Antonio Donato Nobre's warning that if 40 percent of the Amazon rainforest is cleared it could lead to prolonged drought in his 2015 article 'Deforestation and Drought.' *www.nytimes .com/2015/10/11/opinion/sunday/deforestation-and-drought.html*

South Dakota: farmlands watered by rainforests

My first-hand account of the farmlands of the Northern Great Plains in South Dakota come from a visit I made there in September 2016.

The green shields: how rainforests protect people from extreme weather

The loss of mangroves around the coast of Haiti is one of several issues relevant to this part of the book covered on the website of the Critical Ecosystem Partnership Fund. *www.cepf.net/where_we_work/regions/ caribbeanislands/ecosystem_profile/pages/threats.aspx*

The beneficial effects of mangroves for coastal protection are covered in Anna McIvor et al.'s 2012 paper 'Storm Surge Reduction by Mangroves.' *www.conservationgateway.org/ConservationPractices/Marine/ crr/library/Documents/storm-surge-reduction-by-mangroves-report.pdf*

The extent to which deforestation makes the effects of extreme weather more likely is covered by Corey Bradshaw et al.'s 2007 paper 'Global evidence that deforestation amplifies flood risk and severity in the developing world.' *www.onlinelibrary.wiley.com/doi/10.1111/j.1365 -2486.2007.01446.x/abstract*

Wildlife wipe out

The role of scatter-hoarding rodents, especially agoutis, in making up for the lost seed-dispersing activities of now extinct mega-fauna is covered in Patrick Jansen et al, 'Thieving rodents as substitute dispersers of megafaunal seeds.'
www.pnas.org/content/109/31/12610.full.pdf

For a source on the effects of forest fragmentation on biodiversity in the Amazon, see this paper from David Biello
www.scientificamerican.com/article/fragmentation-quickly-des/

CHAPTER 8: PROTEST AND SURVIVE

Mahogany

The history of the campaign on Brazilian mahogany is charted in *Guardians of the Brazilian Amazon Rainforest* by Luiz Barbosa (Earthscan, 2015).

The relationship between logging and total forest clearance is explored in Gregory Asner et al. 'Condition and fate of logged forests in the Brazilian Amazon.' *www.ncbi.nlm.nih.gov/pmc/articles/PMC1538972/*

For the relationship between logging and total deforestation in the Brazilian Amazon, see G.P. Anser et al.'s 2006 paper 'Condition and Fate of Logged Forests in the Brazilian Amazon.'
www.ncbi.nlm.nih.gov/pmc/articles/PMC1538972/

CITES

Art Blundell's 2004 article 'A review of the CITES listing of big-leaf mahogany' provides excellent background. *www.researchgate.net/ publication/231860568_A_review_of_the_CITES_listing_of_big-leaf_ mahogany*

Eating the forests: the Soya Rush

Paulo Adario's 2016 article 'The soy moratorium, ten years on: how one commitment is stopping Amazon destruction' charts the results of the moratorium. *www.greenpeace.org/international/en/news/Blogs/ makingwaves/the-soy-moratorium-10-year-anniversary-stopping-amazon -destruction/blog/57127/*

The extent to which large-scale commercial farming is a factor driving deforestation in different regions is explored in G. Kissinger et al.'s 2012 paper, 'Drivers of Deforestation and Degradation: A Synthesis Report for REDD+ Policymakers.' *www.forestbonds.net/sites/default/files/userfiles/1file/6316-drivers-deforestation-report.pdf*

CHAPTER 9: FOREST DIVIDENDS IN COSTA RICA

First-hand observations are from a research visit I made to Costa Rica during February and March 2016.

The World Economic Forum survey 'Which is the greenest, happiest country in the world?' finds Costa Rica to be top of the list. *www.weforum.org/agenda/2016/07/greenest-happiest-country-in-the-world?utm_content=buffer2d067&utm_medium=social&utm_source=twitter.com&utm_campaign=buffer*

CHAPTER 10: THE BEST FOREST CUSTODIANS

First-hand observations in this chapter are derived from a research visit I made to Asháninka lands in May 2017

Empowerment on the front line

The definitive account of the experiences of the Asháninka at the time of Peru's Shining Path emergency can be found in Friar Mariano Gagnon's *Warriors in Eden* (William Morrow, 1993).

CHAPTER 11: TEMPERATE ZONE RAINFOREST

For a map of the distribution of temperate rainforests see the Forest Legacies website: 'Temperate and Boreal Rainforests of the World: Ecology and Conservation.' *www.forestlegacies.org/images/gis/temperate-rainforests/index.html*

Old World travelers

For an example of research using tiny light-sensitive geolocators, see Susanne Akesson et al.'s 2012 paper, 'Migration Routes and Strategies in a Highly Aerial Migrant, the Common Swift Apus apus, Revealed by Light-Level Geolocators.' *www.journals.plos.org/plosone/article?id=10.1371/journal.pone.0041195*

PART THREE: AFRICA

CHAPTER 12: LAST FRONTIERS: THE CONGO BASIN

Jadvinda Malhi et al.'s 'African rainforests: past, present and future' presents an excellent overview of the African rainforests, including the history of human influence.
www.rstb.royalsocietypublishing.org/content/368/1625/20120312

WWF provides a helpful summary of the Congo Basin forests and people on its website. *www.worldwildlife.org/places/congo-basin*

Congo genocide

BBC correspondent Mark Dummett's 'King Leopold's legacy of DR Congo violence' provides a summary of the brutal regime operating in the Congo Free State. *www.news.bbc.co.uk/1/hi/world/africa/3516965.stm*

An account of atrocities, with testimony from the Rev William Sheppard, was published in the *New York Times* on 5 January 1900.
www.query.nytimes.com/mem/archive-free/pdf?res=9C04E7D9113 CE433A25756C0A9679C946197D6CF

Congo Basin rainforest

For an overview of forest loss in the Congo Basin, see Ernst Celine et al 'National forest cover change in Congo Basin: deforestation, reforestation, degradation and regeneration for the years 1990, 2000 and 2005.' *www.onlinelibrary.wiley.com/doi/10.1111/gcb.12092/abstract*

T. K. Rudel's 2013 paper 'The national determinants of deforestation in sub-Saharan Africa' provides useful background.
www.ncbi.nlm.nih.gov/pubmed/23878341

Assaults on the last frontiers

Nancy L. Harris et al.'s 2017 WRI analysis into deforestation hotspots, 'Using spatial statistics to identify emerging hot spots of forest loss,' includes information on the DRC.
www.iopscience.iop.org/article/10.1088/1748-9326/aa5a2f

For an estimate of new roads to be built up to 2050, see William F. Laurance et al.'s 2014 paper 'A global strategy for road building.'
www.nature.com/nature/journal/v513/n7517/abs/nature13717.html

CHAPTER 13: THE LOST FORESTS OF WEST AFRICA

Claude Martin's book *The Rainforests of West Africa: Ecology—Threats—Conservation* (Birkhäuser, 1991) presents an excellent overview of the state of forests in this region.

Palm oil and cocoa

See W. Gerritsma and M. Wessel's 1997 paper 'Oil palm: domestication achieved?' *www.library.wur.nl/ojs/index.php/njas/article/view/506*

Logging and war

For an example of the effects of illegal logging in Ghana, see Todd Reubold's 2015 article 'New study shines a light on bird loss due to illegal logging in Ghana,' summarizing research published in the journal of biological conservation. *www.ensia.com/notable/new-study-shines-a-light-on-bird-loss-due-to-illegal-logging-in-ghana/*

François Ruf et al present research findings on the ways in which cocoa production has driven rainforest loss in Ivory Coast in 'Climate change, cocoa migrations and deforestation in West Africa: What does the past tell us about the future?' *www.link.springer.com/article/10.1007/s11625-014-0282-4*

Poached forests

The extinction threat posed to hundreds of species of mammals by demand for bushmeat is examined in William J. Ripple et al.'s 2016 paper 'Bushmeat hunting and extinction risk to the world's mammals.' *www.ncbi.nlm.nih.gov/pmc/articles/PMC5098989/*

Ivory Coast poverty traps

My research visit to the cocoa landscapes of western Ivory Coast took place during June 2016.

Raiders of the rainforest

Friends of the Earth's 1992 report into corruption in the Ghanaian forestry sector, 'Plunder of Ghana's rainforests for illegal profit: an exposé of corruption, fraud, and other malpractice in the international timber trade' was used to make a TV documentary.

Structural adjustment

I explore the theme of export-led economic growth as an underlying cause for deforestation. For more, see M. Persson et al.'s 2014 paper, 'Trading Forests: Quantifying the Contribution of Global Commodity Markets to Emissions from Tropical Deforestation' (Centre for Global Development, Working Paper 384).

CHAPTER 14: CLIMATE AND COCOA CHALLENGES

Shortly after I visited Ivory Coast's cocoa landscapes it began to rain again and the drought came to an end. See more in this article by Oiliver Bonner et al on *Bloomberg* (subscription). *www.bloomberg.com/news/articles/2016-07-28/rains-usher-in-relief-for -ivory-coast-cocoa-farms-after-drought-ir6xc22y*

The estimate from the UN FAO that by 2050 food output will need to have increased 70 percent can be found in its 2009 report 'Global agriculture towards 2050.' *www.fao.org/fileadmin/templates/wsfs/docs/ Issues_papers/HLEF2050_Global_Agriculture.pdf*

Power to the people

Emmanuel Bekoe and Fredrick Logah's 2012 paper analyzes 'The Impact of Droughts and Climate Change on Electricity Generation in Ghana.' *www.m-hikari.com/es/es2013/es1-4-2013/bekoeES1-4-2013.pdf*

More on the food, energy and water nexus as an idea that can shape policy can be found in the 2016 UK Parliamentary Paper 'The Water–Energy–Food Nexus.' *www.researchbriefings.files.parliament.uk/ documents/POST-PN-0543/POST-PN-0543.pdf*

The extent to which rising demand for food energy and water is an issue in the context of climate change and other megatrends is explorted in my book *What's really happening to our planet? The facts simply explained* (Dorling Kindersley, 2016).

Helping the cocoa communities

More on Mondelēz's Cocoa Life can be found at their website, *www.cocoalife.org/*

The role of government

More on Ivory Coast's REDD plan can be found in a 2017 Climate Policy Initiative paper by Angela Falconer, 'The Landscape of REDD+ Aligned Finance in Côte d'Ivoire.' *www.climatepolicyinitiative.org/ publication/landscape-redd-aligned-finance-cote-divoire/*

PART FOUR: ASIA AND THE PACIFIC

CHAPTER 15: A SHORT TOUR OF THE EASTERN FORESTS

Jason Daley's 2016 article 'Laser Scans Reveal Massive Khmer Cities Hidden in the Cambodian Jungle' explores recent research on the medieval Khmer empire. *www.smithsonianmag.com/smart-news/laser-scans -reveal-massive-khmer-cities-hidden-cambodian-jungle-180959395/*

Asian forest foods

More on the origin of tea in a 2012 paper by Long Zhang et al, 'Determination of the geographical origin of Chinese teas based on stable carbon and nitrogen isotope ratios.' *www.ncbi.nlm.nih.gov/pmc/articles/PMC3468825/*

The origin of citrus fruit species being identified as Australia is examined by Joaquin Dopazo et al in their 2015 paper 'Evolutionary history of citrus revealed by most comprehensive study to date.' *www.sciencedaily.com/releases/2015/04/150414212311.htm*

For background on the domestication of bananas in Papua New Guinea and nearby islands, see Carol J. Lentfer's 2009 paper, 'Tracing Domestication and Cultivation of Bananas from Phytoliths: An update from Papua New Guinea.' *www.hl-128-171-57-22.library.manoa .hawaii.edu/bitstream/10125/12518/1/i1547-3465-07-247.pdf*

Rice, timber and palm oil

According to the 2003 paper by Plinio Sist et al, 'Sustainable Cutting Cycle and Yields in a Lowland Mixed Dipterocarp Forest of Borneo,' it is possible to log a dipterocarp forest sustainably, so long as simple rules are applied. *www.afs-journal.org/articles/forest/abs/2003/08/F3807/F3807.html*

For a wider discussion about the impacts of logging, see Mongabay founder Rhett Butler's 2012 article, 'Rainforest Logging.' *www.rainforests.mongabay.com/0807.htm*

The story of how Japan's demand for wood affected the Asian rainforests can be found in F. Nectoux and Y. Kuroda, *Timber from the South Seas: An Analysis of Japan's Tropical Timber Trade and Its Environmental Impact* (Gland, Switzerland: WWF International, 1989).

For an overview of what's happening to the rainforests of Indochina, see Yale's Daniel Drollette's 2013 overview 'A Plague of Deforestation Sweeps Across Southeast Asia.' *www.e360.yale.edu/features/a_plague_of_deforestation_sweeps_across_southeast_asia*

For a more specific view on the situation in the Mekong river catchment, see the 2013 WWF report, 'Ecosystems in the Greater Mekong: Current Status, Past Trends, Possible Futures.' *www.awsassets.panda.org/downloads/greater_mekong_ecosystems_report_020513.pdf*

The rapidity of peat forest loss in Indonesia is documented by Jukka Miettinen et al.'s 2011 paper, 'Two decades of destruction in Southeast Asia's peat swamp forests.' *www.onlinelibrary.wiley.com/doi/10.1890/100236/full*

Borneo: the Penan

I mention in relation to the story of the Penan people the very rapid loss of rainforests in Sarawak. For source material on that, see J. E. Bryan et al.'s 2013 paper, 'Extreme Differences in Forest Degradation in Borneo: Comparing Practises in Sarawak, Sabah and Brunei.' *www.journals.plos.org/plosone/article?id=10.1371/journal.pone.0069679*

Transmigration

Rainforest expert Philip Fearnside's 1997 paper, 'Transmigration in Indonesia: Lessons from Its Environmental and Social Impacts,' concludes that 'Indonesia's transmigration program . . . has high environmental, social, and financial costs, while doing little towards relieving population pressure on Java. Transmigration has been an important cause of forest loss.' *www.link.springer.com/article/10.1007/s002679900049*

Thailand and Cambodia

The Global Witness investigation into the links between civil war and the pressures on Cambodia's forests can be found on their website. *www.globalwitness.org/sites/default/files/pdfs/forests_famine_and_war_the_key_to_cambodias_future.htm*

Queensland deforestation

The history and extent of deforestation in Australia is covered in a 2012 paper by Corey Bradshaw, 'Little left to lose: deforestation and forest degradation in Australia since European colonization.' *www.academic.oup.com/jpe/article/5/1/109/1294916*

New Guinea land rights

For an overview of rainforest conservation in Papua New Guinea, see a 2013 CIFOR paper by Andrea Babon and Gae Yansom Gowae, 'The Context of REDD+ in Papua New Guinea: drivers, agents and institutions.' *www.cifor.org/publications/pdf_files/OccPapers/OP-89.pdf*

More on the expedition to the Foja mountains can be found in Stefan Lovgren's 2006 piece in *National Geographic*: '"Lost World" Found in Indonesia Is Trove of New Species.' *www.news.nationalgeographic.com/news/2006/02/0207_060207_new_species.html*

Regarding the Conservation International expedition to the mountains of New Britain, see Jeremy Hance's 2010 article '200 new species discovered in 60-day expedition in New Guinea.' *www.news.mongabay.com/2010/10/photos-200-new-species-discovered-in-60-day-expedition-in-new-guinea/*

Threats to the land rights of tribal peoples in New Guinea are covered in Jeremy Hance's 2010 article, 'Papua New Guinea strips communal land rights protections, opening door to big business.' *www.news.mongabay.com/2010/06/papua-new-guinea-strips-communal-land-rights-protections-opening-door-to-big-business/*

CHAPTER 16: HOW TO DESTROY A RAINFOREST

Ashes to ashes

My eyewitness accounts are from six separate visits to Sumatra and Java between 2013 and 2017. The fire was in Jambi province in Aug 2015.

The estimate that daily emissions from forest fires in Indonesia in 2015 were greater than the daily emissions of the entire United States is from data presented by the Global Fire Emissions Database *www.globalfiredata.org/* and interpreted by the World Resources Institute *www.wri.org/blog/2015/10/indonesia%E2%80%99s-fire-outbreaks-producing-more-daily-emissions-entire-us-economy*

Research into how smoke can reduce rainfall can be found in M. O. Andreae et al.'s 2004 article, 'Smoking rain clouds over the Amazon.' *www.ncbi.nlm.nih.gov/pubmed/14988556*

The World Bank and IMF destruction of forests

On the causes of the 1997 Asian financial crisis, see Robert Wade's 1998 paper, 'The Asian debt-and-development crisis of 1997–?: causes and consequences.' *www.sciencedirect.com/science/article/pii/S0305750X98000709*

How panic led to the Asian financial crisis is documented by Steven Radelet and Jeffrey Sachs in their 1998 paper, 'The Onset of the East Asian Financial Crisis.' *www.nber.org/papers/w6680.pdf*

Jason Tockman explores the connections between IMF loans and conditions and deforestation in his 2001 paper, 'The IMF—Funding deforestation—How International Monetary Fund loans and policies are responsible for global forest loss.' *www.wrm.org.uy/oldsite/actors/IMF/Jason.doc*

For an impression of the dramatic fluctuation in the price of commodities that can accompany financial crisis, see this link to the changing price of palm oil: *www.indexmundi.com/commodities/?commodity=palm-oil&months=300*

Multinationals, paper tigers and greasy palms

Ed Matthew and Jan Willem van Gelder's 2001 report 'Paper tigers hidden dragons' was for Friends of the Earth. *www.foe.co.uk/sites/default/files/downloads/paper_tiger_hidden_dragons.pdf*

Eric Wakker's 2005 Friends of the Earth report is 'Greasy palms: The social and ecological impacts of large-scale oil palm.' *www.foe.co.uk/sites/default/files/downloads/greasy_palms_impacts.pdf*

Friends of the Earth also produced a more focused exposé on the threats posed to the orangutan by the expansion of palm oil, Helen Buckland's 'Oil for ape scandal.' *www.foe.co.uk/sites/default/files/downloads/oil_for_ape_summary.pdf*

CHAPTER 17: TIGERS, PANGOLINS AND HONEYCREEPERS

Snared

See the 2017 article on World Atlas, 'What Is Red Ivory, And How Is Its Trade Threatening The Helmeted Hornbill? *www.worldatlas.com/articles/what-is-red-ivory-and-how-is-its-trade-threatening-the-helmeted-hornbill.html*

Ben Block's article 'Illegal Pangolin Trade Threatens Rare Species' has more on this illegal wildlife trade. *www.worldwatch.org/node/6198*

Pacific aliens

C. E. Asian et al 'Imperfect replacement of native species by non-native species as pollinators of endemic Hawaiian plants' explains how the introduced birds now found on the Hawaiian islands do not fulfil the same pollination functions as the native species. *www.ncbi.nlm.nih.gov/pubmed/24372761*

The Western Ghats and other Asian hotspots

Kamal S. Bawa et al.'s 'Western Ghats Biodiversity Hotspot' is a useful paper from the Critical Ecosystems Partnership Fund. *www.cepf.net/Documents/final.westernghatssrilanka_westernghats.ep.pdf*

The WWF also provides an overview of the status of forests in the Western Ghats. *www.wwf.panda.org/what_we_do/where_we_work/western_ghats/*

CHAPTER 18: POACHERS INTO GAMEKEEPERS?

The proposal to convert the legal status of former logging concessions into permanent forest areas is discussed in Rhett A. Butler's 2013 article, 'Indonesia should convert logging concessions to protected areas to stop deforestation for plantations, argues study.' *www.news.mongabay.com/2013/09/indonesia-should-convert-logging-concessions-to-protected-areas-to-stop-deforestation-for-plantations-argues-study/*

PART FIVE: WORTH MORE ALIVE THAN DEAD

CHAPTER 19: GLOBALIZED DEFORESTATION . . .

For more background on why WTO rules were seen as an engine for further forest loss, see John W. Friede's paper, 'Global Trade and the Rainforests: Corporate Growth vs Indigenous Prosperity in Tropical Countries.' *www.environment.yale.edu/publication-series/documents/downloads/0-9/98friede.pdf*

CHAPTER 20: . . . AND GLOBAL SOLUTIONS

For more on Andrew Mitchell's estimate that about $135 billion is invested each year in activities that damage the tropical rainforests, see the Global Canopy press release. *www.globalcanopy.org/new-analysis-and-ranking-c-135-billion-export-trade-forest-risk-commodities-continues-to-destroy*

The Prince's Rainforests Project

The Prince's Rainforests Project's proposals for funding to halt deforestation are set out in the report, 'An Emergency Package for Tropical Rainforests,' presented to G20 leaders in London in April 2009. *www.princeofwales.gov.uk/sites/default/files/documents/Report%20-%20March%202009.pdf*

Copenhagen and beyond

Daniel J. Weiss and Andrew Light's article in *Grist*, 'What you need to know following the Copenhagen climate summit,' summed up the outcomes of the summit; the money pledged for forests is in the 'Major Financial Commitments' section. *www.grist.org/article/2009-12-23-what-you-need-to-know-following-copenhagen-climate-summit/*

Paris 2015: a turning point?

The statement 'Unlocking the Potential of Forests and Land Use' sets out the intention of the UK, Norway and Germany to spend an additional $5 billion on tropical forest conservation. *www.bmub.bund.de/fileadmin/Daten_BMU/Download_PDF/Klimaschutz/joint_statement_redd_cop21_en_bf.pdf*

Brazil's national plan adopted at Paris can be viewed here: *www4.unfccc*
.int/submissions/INDC/Published%20Documents/Brazil/1/BRAZIL%20
iNDC%20english%20FINAL.pdf

Indonesia's national plan adopted at Paris can be viewed here:
www4.unfccc.int/submissions/INDC/Published%20Documents/
Indonesia/1/INDC_REPUBLIC%20OF%20INDONESIA.pdf

Mike Gaworecki's article 'New research shows why forests are
absolutely essential to meeting Paris Climate Agreement goals' is a fine
summary of the importance of forests in meeting the aims of the Paris
Agreement. *www.news.mongabay.com/2017/11/new-research-shows-why*
-forests-are-absolutely-essential-to-meeting-paris-climate-agreement-goals/

CHAPTER 21: VALUING NATURE AND RAINFOREST

The value of nature

Earth Time-Lapse can be viewed here:
www.earthengine.google.com/timelapse/

Oliver Milman and Stuart Leavenworth's article 'China's plan to cut
meat consumption by 50 percent cheered by climate campaigners'
appeared in *The Guardian. www.theguardian.com/world/2016/jun/20/*
chinas-meat-consumption-climate-change

Possible connections between climate change, drought and conflict in
Syria are examined by Colin P. Kelley et al in their 2014 paper, 'Climate
change in the Fertile Crescent and implications of the recent Syrian
drought.' *www.pnas.org/content/112/11/3241.short*

CHAPTER 22: FUTURE FORESTS

A little more background on and history of Tanzania's Amani Forest can
be found in a pamphlet published by the Tropical Biology Association
'Amani Forest Nature Reserve—An Introduction.' *www.tropical-biology*
.org/wp-content/uploads/2015/01/AmaniNR_FINAL.pdf

Zero deforestation companies

The Tropical Forest Alliance is a large country and company initiative
geared to protecting tropical forests, including via zero deforestation
supply chains. *www.tfa2020.org/en/*

For Tropical Forest Alliance members' progress toward meeting zero deforestation goals, see Charlotte Streck's 2020 report. *www.tfa2020.org/en/go-private-sector-progress-zero-deforestation -commitments-really-means/*

Nowhere to hide

For more on the very impressive work of Global Forest Watch, take a look at their website: *www.globalforestwatch.org/*

The Bonn Challenge: forest restoration

For more on the Bonn Challenge on forest restoration, see their website: *www.bonnchallenge.org/*

Protect and conserve

On the importance of protected areas in averting the extinction of rainforest wildlife, the 2012 paper by William F. Laurance et al 'Averting biodiversity collapse in tropical forest protected areas' sets out the issues very well. *www.nature.com/articles/nature11318*

Jonas Geldman et al.'s 2013 paper, 'Effectiveness of terrestrial protected areas in reducing habitat loss and population declines,' underlines the vital roles played by National Parks and other protected areas. *www.sciencedirect.com/science/article/pii/S0006320713000670*

Jonas Geldman et al chart the rising pressures on protected areas in their 2014 paper, 'Mapping Change in Human Pressure Globally on Land and within Protected Areas.' *www.onlinelibrary.wiley.com/doi/10.1111/cobi.12332/abstract*

For the role of protected tropical forest areas in holding carbon, see the 2010 paper from Jörn P. W. Scharlemann et al, 'Securing tropical forest carbon: the contribution of protected areas to REDD.' *www.macroecointern.dk/pdf-reprints/Scharlemann_O_2010.pdf*

Offset

In 2016 a group of leading conservation groups published a paper calling for a linkage between charging emissions caused by aviation and funding to conserve tropical forests. Their paper, 'Linking Flight and Forests: The Essential Role of Forests in Supporting Global Aviation's

Response to Climate Change,' can be found here: *www.globalcanopy
.org/sites/default/files/documents/resources/Linking%20Flight%20and%20
Forests%20Briefing%20Paper%20Apr%202016%20FINAL.pdf*

For more on the work of Fauna and Flora International's work in
Liberia, see the organization's website: *www.fauna-flora.org/countries/
liberia*

Big ideas and forest connections

For a fuller explanation of Martin von Hildebrand's proposals for a
major network of connected protected areas across the Amazon, see
Todd Reubold's interview. *www.ensia.com/interviews/martin-von
-hildebrand-an-audacious-plan-for-the-amazon/*

Toward the end of this chapter I express a sense of optimism. For
a supporting view on that, see Doug Boucher et al.'s 2014 paper,
'Deforestation success stories—Tropical Nations Where Forest
Protection and Rainforest Policies Have Worked' for the Union of
Concerned Scientists. *www.ucsusa.org/sites/default/files/legacy/assets/
documents/global_warming/deforestation-success-stories-2014.pdf.*

NOTE ON THE MAPS

The three area maps in this book present an approximation of historical and current tropical rainforest cover. Note that at the scale presented in this book, small areas of forest and forest-farm mosaics are not visible. Areas of logged, degraded or fragmented forest are also not distinguishable from primary largely undisturbed tropical rainforest. What is visible, however, is the approximate original extent of equatorial rainforests and where the largest mainly contiguous blocks remain. For greater detail, you can view the maps online at *www.tonyjuniper.com*

Historical forest cover is taken from maps presented in *The Last Rain Forests: A World Conservation Atlas* (Oxford University Press, 1990). *Grid Arundel 2015* is the source for current rainforest cover in Latin America, 'Tropical Forest in Latin America' (*www.grida.no/resources/6941*). For current tropical forest cover in Africa, we drew on the map based on MODIS data presented by Mongabay (*www.rainforests.mongabay.com/congo/*). The current rainforest cover for Asia and the Pacific is also taken from *Grid Arundel 2015*, 'Tropical Forest in Southeast Asia and Oceania' (*www.grida.no/resources/6957*).

PHOTO CREDITS

The color photos in this book were taken by **Thomas Marent**, who has been photographing the rainforest for the past twenty-five years. He is best known for his book *Rainforest: A Photographic Journey* (Dorling Kindersley, 2006), which has appeared in fifteen languages. Selections of his rainforest work can be seen on his website, *www.thomasmarent.com*

Many of the black and white images in the book are also by Thomas Marent (TM), or by the author (TJ). We are also grateful to the following for photos used in the main text:

WikiCommons: p.50 (Terry Hughes); p.123; p.129; p.231 (Max Chiswick); p.235 (Medicaster); p.261 (ZSM); p.280 (Francesco Veronesi); p.284.

NASA: p.57; p.145.

Manuel Arroyo-Kalin: p.111.

Gleilson Miranda/FUNAI/Survival: p.123.

Greenpeace: p.135; p.165; p.168; p.299; p.308; p.313; p.324.

Getty Images: p.140 (Marcos Alves/Moment Open); p.225 (Cultura RM Exclusive/Philip Lee Harvey); p.320 (Christopher ScottGallo Images); p.324 (Mint Images/Frans Lanting); p.355 (AFP); p.359 (Anwar Hussein/ WireImage); p.367 (François Guillot/AFP); p.394 (Cesar Carrion/AFP).

Friends of the Earth: p.155; p.159; p.350.

Cool Earth: p.183 (Alicia Fox); p.191 (Mark Ellingham); p.192; p.196 (Alicia Fox); p.295; p.382.

Cocoa Life: p.269.

Bruno Manser/ Global Witness: p.287.

UNESCO: p.338.

Rainforest Action Network: p.346.

Charles Clover: p.362.

Google Earth Time-Lapse: p.374.

Global Forest Watch: p.381.

Flora & Fauna International: p.390.

ACKNOWLEDGMENTS

A great many people have contributed to my understanding and experience of the subjects covered in this book, not least my many Friends of the Earth colleagues that it was my privilege to work alongside during the 1990s and 2000s, including Charles Secrett, Andrew Lees, Simon Counsell, Roberto Smeraldi, Tim Rice, Adrian Barnett and Ronnie Hall. I am similarly indebted to my friends and colleagues at the Prince's Rainforests Project (PRP), and more latterly the Prince of Wales's International Sustainability Unit (ISU), from whom it has been my good fortune to gain many insights, ideas and research leads, including Edward Davey, Justin Mundy, Jack Gibbs, Sir Graham Wynne and Beatriz Luraschi. His Royal Highness The Prince of Wales has also been a source of great inspiration, not least in relation to his consistent and wise advocacy regarding the central role of agroforestry in building integrated landscape-scale solutions to deforestation.

I would like to express gratitude for the advice and assistance of John Sauven at Greenpeace regarding that organization's campaigns for the forests, and Mike Parr of the American Bird Conservancy for his input on the threatened endemic birds of Hawaii. Neil Burgess of the UN Environment's World Conservation Monitoring Centre provided much valuable input, including in relation to the recent science base. Aida Greenbury of Asia Pulp and Paper (APP) and her colleague Dolly Priatna assisted in my understanding of the complex issues linked with deforestation in Indonesia and I'd like to thank my friend and Robertsbridge colleague Brendan May for his company during our work assisting APP on its journey toward sustainable business.

I was very pleased to have the help of Mondelēz in ensuring both an informative and safe visit to the cocoa landscapes of Western Ivory Coast, including Jonathan Horrell, Cathy Pieters and Mbalo Ndiaye and in the field Mohammed Amin, who guided me to the villages and communities I met, and local teacher Gueu Mande, who provided excellent translation during those travels. I am grateful to Jean-Paul Aka of Ivory Coast's REDD agency for explaining his work to me.

Professor Edward Wilson was kind enough to explain his ideas about island biogeography and it was my privilege to have the scholarly advice of Dr John Hemming in both advising on key issues and also looking at some of the draft text. Andrew Mitchell of the Global Canopy Programme was similarly an invaluable source of advice.

I am enormously grateful to Matthew Owen, Chris Kuahara and Lewis Gillingham of Cool Earth for their assistance in getting me safely into Asháninka lands in the rainforests of the Peruvian Amazon and for the additional company of Cláudio Cardoso, who joined us on that trip. On the subject of indigenous forest peoples. I am grateful to Martin von Hildebrand, who kindly explained his work in that sphere in Colombia, and also his vision for a connected Amazon-wide corridor of protected forest. I was grateful to have my knowledge of the coastal forests of central Chile expanded via a couple of trips into those ecosystems with Chilean conservationist Max Correa.

I would like to thank Crystal Davis for her help in explaining the Global Forest Watch initiative and to Peter Bunyard and Deborah Lawrence for so clearly setting out their views and insights regarding the role of tropical rainforests in the global water cycle. Andrew Parker and Tom Prescott provided me with input on the value of biodiversity (in relation to biomimicry and medical innovation respectively). Maddie Juniper supplied excellent research support and I am indebted to Jonas Geldman of the University of Cambridge, who helped me with technical sources regarding the role of protected areas. I was pleased to speak with Laura Fox and Zoe Cullen of Fauna and Flora International (FFI), who told me of their perspectives on conservation finance.

Because of the great breadth of complex subject matter and the multiple themes that needed to be combined, this was a tricky book to write. If I have managed to rise to the challenge of producing an interesting and accessible narrative, then that is in large part down to Mark Ellingham, my patient and determined editor at Profile Books, who expertly guided the text through its several drafts and to what I hope readers will find an interesting and informative read. I am grateful to my agent Caroline Michel at Peters Fraser & Dunlop, both for initial encouragement in developing the original idea and also for making arrangements with Profile Books to publish it.

On the color pictures, I'd like to record the great pleasure of working with Thomas Marent's wonderful images. Thanks also to Dominic Beddow, who devoted great effort to producing the excellent maps that depict forest cover across the three main regions. This was a difficult task, not least because of the challenge of matching such a vast amount of spatial detail into such a small space. I am grateful, too, to Henry Iles for page design and turning green images into black and white, Nikky Twyman for proofreading, Peter Dyer who did the cover, and Bill Johncocks, who prepared the fine index.

Last but not least I'd like to thank my wife Sue Sparkes for help and support throughout, including accompanying me during field research in the Western Ghats in India and to the rainforests of Costa Rica.

INDEX

Note: The index covers the text and illustration captions but not the color Plates sections, endnotes or image credits. Page *references in italics* indicate a relevant illustration separated from any coverage in the text.

A

Abidjan 237–8, 242, 257, 259, 271
Adoulaye, Diarrassouba 245, 264
aerosols 21
Africa
 ancient civilizations 220
 map 218–19
 slavery effects on forest cover 222–3
Africa, Central
 deforestation rate 227
 indigenous people 201–2, 230–3
Africa, West
 deforestation 234, 236–9, 243–4, 257–8
 droughts 256
 Guinea forests 101, 118
 hunting 240
agoutis 70, 113, 152–3, 196
agriculture / farming
 commercial, as a deforestation driver 136–7, 181
 industrial scale 141
 pre-Columbian civilizations 108, 110–11
 prehistoric African 220
 by settlers 188, 254
 subsistence farming 92, 186, 193, 240, 242–4
 unsuitability of cleared rainforest 58
 see also smallholders
agroforestry
 bananas 266
 coffee 340–1
 methods 193, 266–7, 273, 340–1
 promoting 391, 394
Air Artists 158–9
air pollution, from forest fires 302
air quality benefits 176
air resistance, overcoming 97
aircraft emissions 390–1, 393

aircraft surveillance 23, 42, 146, 169, 333, 385
Aka, Jean-Paul 271–4
Akasombo Dam 260, 261
albedo 35–6
alien species, imported 322–7
altitude effects 83, 88
Amazon River
 daily discharge volumes 20
 exploration 110
 seasonal depth variation 19
 seed dispersal by fish 69
Amazon basin
 cloud formation 15, *17*
 droughts 45
 eco-cultural mosaics scheme 395–6
 as a freshwater system 19–21
 history of forest clearance 132–7, 156, *376, 377*
 Maraca Project 60, 62
 nutrients from Saharan dust 56
 planetary significance 23–8, 29
 population collapse 10, 118
 rainfall trends 28, 45
 soya threat 164–70
 varzea flooding 20, 113
 weight of water vapor emitted 22
The Amazon Fund 360
Amazon Watershed,
 by George Monbiot 157
Amazon Working Group 156
The Americas
 endangered animals 209
 forest clearances 126–34
 land bridge between 77, 81, 215
 map 104–5
 pre-Columbian civilizations 106, 108
 temperate rainforests 204, 209
Amin, Mohamed 267–8
Andean civilizations 117

animal feed, soya for 135, 165
animals
    endangered South American 209
    extinct megafauna 152, 210
    protecting trees 71–2
    seed dispersal by 66, 68–74, 152–3, 177
    vulnerability of isolated populations 151
Antarctica 37, 208, 210
    British Antarctic Survey 37–8, 53
antbirds and antshrikes 86
anti-reflective coatings 96
ants 71–3, 83, 86, 97
apes
    howler monkey comparison 75
    orang utan 314, 321, 322
    status of surviving great apes 321–2
    see also gorillas
APP (Asia Pulp and Paper) 311–14,
331–7, 339–40, 342, 386
    Belantara Foundation 337
    Forest Conservation Policy 332, 337
aracaris 69, 71
Araucaria trees 76
'arc of deforestation' 133, 138, 377
Aripuanã River massacre 121
armadillos 77, 210
armed forces, abolition 180–1
Asabliko, Ivory Coast 254–5
Asháninka Communal Reserve 185
Asháninka people 122, 182–99, 340,
384, 395
Asia and the Pacific
    ancient and modern forest cover 278
    map 276–7
Asian financial crisis 307, 310, 312,
345
Atlantic forests, South America 73–4,
126–7, 138
the atmosphere
    interaction with ecosystems 53
    rainforest contribution to circulation 26
Australia
    Great Barrier Reef 49, 52, 293
    Queensland deforestation 293
    rabbit proof fence 18
aviation sector 390–4
avocados 102, 109, 112
Awajún people 199

B

Ba'Aka people 230, 231
Bali, UN Climate Change Conference,
2007 356–9, 362

Ban Ki-moon 357, 369
bananas
    in agroforestry 266
    Costa Rican exports 171, 176
    origins 282–3
    pollination by bats 67
    pre-Columbian cultivation 112
Barbie, Ken and, banner 313, 314
Barco, Virgilio 200
Barro Colorado (island) 150, 152
bats
    Honduran tent-making bats 72, 72
    as pollinators 67
    seed dispersal by 68, 72
    variety in Costa Rica 81, 84
bee-eaters 75, 316
beef see cattle
Belantara Foundation (APP) 337
Benin 234, 238, 260
Berlusconi, Silvio 362
big-leaf mahogany 155, 161
bigleaf maple 206
biochar 112
biodiversity
    and food sources 194
    and geographical isolation 80–1
    Hawaii 323–5
    Indochina 292
    Leuser landscape, Sumatra 321
    mega-diversity 79–83
    Tai Forest National Park 246
    Tanzania 100, 252, 380
biodiversity hotspots 99–102, 149,
152, 327
biofuels
    corn ethanol 143–4
    from palm oil 286
    from sugar cane 282
bioluminescence 87
biomass, vertebrate 373
biomimetic research 94–5
the biosphere concept 53
biosphere reserves 335, 338
biotic pump theory 25–6, 28–9, 44,
46, 48–9, 53
birds
    Central American endemics 89
    endangered 310
    migratory 84, 211–14
    pollination by 66, 67, 324, 326
    seed dispersal by 66, 68, 177, 318
    unique to Hawaii 323, 324
    variety in Costa Rica 81, 84
    see also individual species

Bodélé Depression 56
Bolivia
  opposition to CITES listing for
  mahogany 163
  pre-Columbian earthworks 110
The Bonn Challenge 386–7
books
  Amazon Watershed, by George Monbiot
  157
  The Heart of Darkness, by Joseph Conrad
  224
  The Putumayo, the devil's paradise . . . by
  Walter Hardenberg *120*
  In the Rainforest, by Catherine Caufield
  122
  The Tropical Rain Forest, by Paul Richards
  17
Borneo
  APP target landscapes 333, 339
  drought and El Niño 45
  effect of pulpwood operations 310–11
  effect of timber operations 284–6, 291
  emissions from burning peat 45, 302,
  377
  fish as seed dispersers 69
  indigenous people 287–9
  peat swampland 41
  scale of deforestation 286, 291, 302,
  327
  Tanjung Puting National Park *322*
  Tawau Hills National Park 82
Brazil
  action on deforestation 361, 370
  carbon storage and carbon emissions 41
  colonization 117–18
  crackdown on mahogany exports 163–4
  deforestation rates 137, 170
  discovery of 'lost cities' 110
  drought of 2014 138
  highways, BR163 and BR364 133, 164
  opposition to CITES listing for
  mahogany 163
  political inconsistency 378
  see also Amazon basin
Brazil nut trees 70
Britain
  biodiversity, compared with Costa Rica
  81
  Ghanaian economic reforms 252
  illegal timber imports 154, 156–7,
  159–60, 166–7
  lacking a conservation strategy 175
  Saharan dust 55
  Ty Canol woodland 214–16
British Antarctic Survey (BAS) 37, 53
British East India Company 280

bromeliads 33–4, 59, 86, 88, 91, 211
  absence in West Africa 246
  pineapples as 109
Brown, Gordon 362, 365
building design improvements 97
Bunyard, Peter 24–8
Burgess, Neil 100–1, 252, 380
'bushmeat' 196, 226, 241
business case for forest protection 263,
267, 269–71
Bustamante, Adelaida 198–9
butterflies
  Colombian 94–6
  Costa Rica 86
  iridescence 94
  Peruvian 81
  as pollinators 65

C

caching seeds 73, 152
calima conditions 56
Cambodia 185, 278, 283, 291–2
camera traps 92, 195–8, 384
Cameroon 224, 238–9
camouflage 94
campaigning targets 159, 167
Canary Islands 56, 57
the canopy
  benefits to cocoa 266
  benefits to coffee 341
  height 81–2
  relative humidity under 27
carbon
  black carbon from fires 302
  release by drying rainforests 45
  storage by rainforests 37, 39–42, 74,
  336
  storage using agroforestry methods 341
carbon charges 390–1
carbon density 47
carbon dioxide
  capture by forests 42–3
  deforestation contribution 42–3
  funding emission reductions 272–3
  generation by forest fires 43, 45, 302
  ice core records 38, 222
  post industrialization 38–9
carbon offsets 361, 390–5
Cardoso, Cláudio 197
Care (NGO) 264–5, 267–8
Cargill Inc. 164–9, 366
Caribbean
  biodiversity 147, 149
  Columbian landfall 106

coral reefs 49, 52
  deforestation 131, 156
  hurricane damage 144–6
  vulnerability to logging 131
Caribbean Plate 77
Carnegie Mellon University 372, 375
Carvajal, Gaspar de 110
catfish, migratory 20–1
Catholicism 115, 122
cattle ranching and deforestation 126–34, 169–70, 172–4
Caufield, Catherine 122
Cecropia trees 71
Central African Republic 224, 231
cerrado woodlands 134
Chad 56
chainsaw, inflatable 158–9
charcoal
  production and tree losses 126–7, 146, 210, 227–8
  use by Iron Age Africans 221
  use by pre-Columbian farmers 112
Charles, Prince see Prince of Wales
chickens
  domestication and contemporary importance 280, 281
  kept by the Asháninka 193–5
  linked to deforestation 167, 168
children
  abductions 200
  Asháninka people 196, 197
  infant mortality 192–3
  in National Parks 180
Chile 208–11
China
  fossil fuel use 373
  meat consumption guidelines 379
  satellite evidence 378
Chinese medicine 319–20
Chiribiquete National Park 396
chocolate
  message to consumers 268–9
  origins 108
  traceability 268
  yields of cocoa 254, 265
  see also cocoa
chocolate manufacturers 244, 263–4, 266–7, 270, 340, 381, 394
CITES (Convention on the International Trade in Endangered Species) 160–4, 312
  African, demands on the forests 226
  forest clearance and urbanization 127, 238, 270, 387

'lost cities' 106, 110–15, 278
  rapid expansion of Asian cities 297, 307
citrus fruits 282
Clarita (sister-in-law of Max Correa) 210
climate, global, links to forests 16, 35–6
climate change
  aircraft emissions and 393
  effect on cocoa production 256, 272
  emergence as a political concern 353–4
  feedback effects through forests 43–4
  as a funding priority 272–3
  links to defaunation 73–4
  links to deforestation 11, 39, 47–9
  Paris Summit, 2015 337, 365, 367–71, 386, 394
  reduced rainfall 262
  slowing through reforestation 356, 388, 395, 397
  vulnerability of isolated populations 151
  see also global warming
Climate Change Act, 2008 367
Climate Change Conferences (UN)
  Bali, 2007 356–9, 362
  Copenhagen, 2009 362, 363–6, 394
  Paris, 2015 337, 365, 366–71, 386, 394
Clinton, Hillary 362
cloud forests
  climate change vulnerability 151, 326
  Colombian 96
  Cost Rican 33, 34, 91–3, 173
  as a phenomenon 33–5
  Western Ghats 328
  see also fog
cloud formation / cloud seeding 15, 17, 21–2, 24, 27
clouded leopards 304, 340
coca / cocaine 133, 185–6, 190
cocoa
  assistance to farming communities 263–73, 340
  bats as pollinators 67
  benefits to indigenous people 189, 193, 197
  history and status of cultivation 108, 256
  pest control 99
  plantation driving deforestation 127, 133, 150
  responsible sourcing 381
  utility of wild genes 102
  in West Africa 235, 236–7, 243–7, 254–7, 263–7
  yields 254, 265, 272–3
  see also chocolate

Cocoa Life initiative 255, 263–4, 267, 269–70
coffee
    agroforestry initiatives 340–1
    forest clearance for 126, 128, 130–1, 146
    indigenous farmers 189–90, 193–5, 197
    utility of wild genes 102
    Vietnam 292
    Western Ghats 328
Colombia
    Chiribiquete National Park 396
    cloud forests 96
    eco-cultural mosaics proposal 395–6
    extent of deforestation 132–3
    policies toward indigenous peoples 199–203
    rainforest biomimicry 94–6
colonization
    Americas 108, 116–17, 127, 131, 184
    Asia 279–80
    Brazil 117–18
    Congo Basin 220–1
    Hawaii 325
    and national elites 232
colonnade formation 205
Columbus, Christopher 106–7, 115–16, 118, 131, 171, 279, 282
composting 111–12, 194
concessions
    logging 305–6
    mapping 337–40
    palm oil 305
    wildlife assessment 336
condensation
    and air-flow 26
    latent heat of 25
    transpired water 21–2
conflict
    and logging 237–8, 292
    over access to protected areas 388
    Syrian drought of 2006 and 375
conflicts
    Central America 130, 181
    Congo Basin 224
Congo Basin 220–33
    Cool Earth in 189
    deforestation pressure 32, 226–30
    migratory birds 213
    peat underlying 41
    populations 224, 230–1
    rainforest extent 224
    see also DRC
Congo Free State 223
conifers, giant 205
Conrad, Joseph 224

Conservation International 295–6
conservation research 195–7
convergent evolution 74–5
Cook, James 325
Cool Earth 182–4, 186, 188–90, 192–200, 340
    scope, aims and methods 189, 197–8
Copenhagen, UN Climate Change Conference, 2009 362, 363–6, 394
coral reefs
    bleaching 52–3
    deforestation effects on 49–54
'The Coral Triangle' 51
corn see maize
corn ethanol 143–4
corn syrup 141–2
corporations see multinational companies
Correa, Max 210
corruption
    in Africa 242, 247–52
    exposure, using technology 383
    and illegal logging 156
    in Indonesia 306, 334
    and international aid 349, 388
    in South-East Asia 292–3
Costa Rica
    abolition of armed forces 180
    cloud forests 33, 34, 91–3, 173
    enlightened conservation policies 130, 163, 362
    fruit exports 171, 176, 181
    fungi 58
    increased forest cover 172, 174–5
    La Selva Biological Station 69, 178
    Los Angeles Cloud Forest 33, 93
    lowland rainforests 83–7, 211
    National Parks 82, 83, 85, 172, 174
    Talamanca Mountains 88, 100
    tourism 171–2, 174–9, 181, 393
    tree seed dispersal 70
    upland rainforests 31, 87–91
    variety of wildlife 81
    well-being and life expectancy 180–1
Counsell, Simon 247
Coveja, Peru 192, 195
'crony capitalism' 307
crops
    domestication by indigenous peoples 108–9, 281–3
    genes from wild relatives 102
    genetically modified 141
    pollinated by bats 67
    pollination by rainforest species 99
    the 'three sisters' 112

Cross River Park, Nigeria 238
Cuba 131, 147, 156
cuckoos 213
Cullen, Zoe 391–2
cultural diversity 395, 398
Cuneo, Michele de 115–16
curassows 86
Cutivereni, Peru ('Cuti') 189–90, 198

D

Dahomey Gap 234
Darwin, Charles 209, 283, 324
Davis, Crystal 383–5
Davos, World Economic Forum, 2001
349–53
De Boer, Yvo 357
debt bondage 119, 199–200
debt crisis 251–2
decision makers
    Africa and Latin America contrasted 232
    and the Bonn Challenge 387
    in supported communities 198
decommodification 382
decomposition
    decaying wood 58, 62
    leaf litter 59
    and nutrient recycling 62–3, 92, 205
    peat formation 40–1
    upland forests 82–3
defaunation 73–4
deforestation
    American forest clearances 126–34
    'arc of deforestation' 133, 138, 377
    carbon emissions and 39
    Central Africa, rate of 227
    and commercial agriculture 136–7, 181
    community management and 201
    Congo Basin threats 224, 226–30
    and desertification 25, 27
    effect on corals 49–54
    effect on rainfall 18, 28, 30, 138, 257
    effect on temperature 35–6
    following resettlement schemes 290
    'herring-bone' pattern 133, 376, 377
    illegal clearance in Sumatra 333, 334,
    336
    links to climate change 11, 47–9, 354
    logging as first stage 47–8, 131, 156,
    188, 226, 232, 284–5
    mechanized 127
    preliminary degradation and
    fragmentation 47–9, 133, 376, 377
    Queensland, Australia 293
    remote effects 11, 141

reversal in Cost Rica 172
soya threat 164–70
structural adjustment programs and
252–3
trends, twentieth century 355
trends, twenty-first century 384
West Africa 236–9, 243–4, 257–8
zero deforestation companies 381–82
desertification 25, 27, 43, 78, 141
deserts, neighboring rainforests 27
development goals
    and the 2015 Paris summit 369
    conflict with forest conservation 306
    conflict with indigenous people 233,
    287–9
Dewi, Sonya 340–2
diamonds 224, 228, 237
disease
    bird malaria and bird pox 326
    Ebola virus 221
    susceptibility of indigenous peoples
    116–17, 293
    swollen shoot disease 273
    Zika virus 140
display screens 94
DIY stores, targeting 159, 163
domestication
    of animals 116
    of crops 108–9
Dominican Republic 106, 144, 146, 263
Drake, Francis 107, 109, 363
DRC (Democratic Republic of Congo)
102, 220, 223–4, 225, 226–8, 230
droughts
    Amazon basin 45
    associated with Ice Ages 78
    biotic pump effect 46
    deforestation and 138, 257, 262
    health effects 140
    Indonesia, 2015 300–3, 309
    risk of forest fires 32, 300
    risk to hydroelectric projects 139, 172,
    258–62
    risk to profitability 365–6
    Syria, 2006 375
    West Africa 256
drug traffickers / narco-traficantes 127,
133, 185–6, 189–90, 195
    see also coca
Dutch East India Company 280

E

Earth Summit, Rio, 1992 208, 232,
346, 357, 363

Earth Time-Lapse presentation 375–7
Eating Up the Amazon, report 167
Ebola virus 221
'Ecochickens' demonstration 167, 168
ecological interconnectedness 63–8
economic benefits of conservation
174–5, 181, 273, 355, 359, 388, 394
economic liberalization 307, 345–6
economic recession 138–41, 308
ecosystems
    cerrado woodlands 134
    comparable, convergence in 74–5
    marine 49, 51–2
    see also biodiversity; wildlife
ecotourism 174, 178, 181, 393–4
Ecuador 79, 110, 127, 149, 395
'edge effects' 47–8, 150–1, 225
education
    cocoa farmers 263, 265, 268, 271
    indigenous people 195
El Niño 45
El Salvador 48, 128, 130, 181
elephants
    Asian elephants 328
    as 'mega-gardeners' 68
    Sumatran elephant 318–20, 321
emergent trees 68, 82, 246, 249
endemic species
    of biodiversity hotspots 99–100
    Caribbean 147, 148
    Central America 86, 89
    defined 89
    Tanzania 380
energy, renewable 175–6, 282, 390,
394
energy efficiency 39
energy security 398
environmentalists
    approach from APP 331
    murdered 128, 288
    need for continuous campaigning 397
epiphytes 23, 35, 72, 91, 206
'ethical shoplifting' 158
Ethiopia 102, 370
Europe
    human migration to 242, 257, 265
    timber imports 10, 292, 387
evaporation rate, forests 22–3
evolution
    convergent evolution 74–5
    emergence of flowering plants 76, 80
    founder populations 323
    history of rainforests 76–7
extinct megafauna 152, 210

extreme weather
    deforestation and 48, 144–7
    global warming and 39
    hurricanes and tornadoes 49, 144, 151
    vulnerability of isolated populations 151

F

Fabius, Laurent 368, 369
farming see agriculture
fast food 130, 167–8, 173
Ferrero SpA 340
fertilization
    composting 111–12, 194
    pollution from 286
    pre-Columbian 111–12
    wind-borne 56, 58
FFI (Fauna & Flora International) 391
fig trees 71
Figueres, Christiana 369
Figueres, José Maria 175
financial crisis, Asia 307, 310, 312, 345
fire detection 96
fire risk see forest fires; peat swamps
fireflies 87
firewood 146, 241, 245
fish, seed dispersal by 69
'fish-bone' deforestation 133, 376, 377
flooding reduction by forests 48, 146
flowering plants
    emergence 76, 80
    in sexual reproduction 64, 67
flycatchers 89, 91, 212
fog, moisture from 209–10
    see also cloud forests
Foja Mountains 295–6
food, with rainforest origins 281–3
    see also crops
'food, water and energy nexus' 262
Food and Agriculture Organization
(UN) 137, 258
food security 387, 398
Ford, Henry 132
forest clearance see deforestation
forest conservation funding 355–6,
360–1, 365, 383
Forest Conservation Policy (APP) 332,
337
forest fires
    drought and risk of 32, 300
    peat swamp burning 45, 298–300, 377
forests
    evaporative cooling 84
    management of Amazonian 113

replanting and regeneration 176–7, 180, 210, 272–3, 386
see also rainforest
fossil fuel burning 39, 373, 393
'Fourth Industrial Revolution' theme 374
fragmentation of forests
effect on emissions 47
effect on wildlife 149, 152
Friends of the Earth
Big Ask campaign 363
climate change report, 1990 37
Greasy Palms report 314
iconic frog 92
Mahogany is Murder protests 154, *155*, 156
Paper Tiger, Hidden Dragons report, 2001 311
Public Eye on Davos campaign 351–2
strategy toward multinationals 345, 347, 349
technological resources 386
see also Lees, Andrew; Secrett, Charles
Friends of the Earth Ghana 250
Friends of the Earth International 350, 356–8
Friends of the Earth Switzerland 351, 353
frogs
Costa Rica 34, 81, 87, 91–2, *93*
local variability 100
New Guinea 295–6
Tanzania 380
fruit
exports, Costa Rica 171, 176, 181
rainforest origins 282–3
FSC (Forest Stewardship Council) 160
funding
carbon offsets 361, 390–5
climate change priority 272–3
for forest conservation 355–6, 360–1, 365, 382
from lotteries 394
World Bank, in Ivory Coast 272
fungi
mycorrhizae 62–3, *73*
nutrient recycling *58*, 60
furniture trade 155–6

G

G20 meeting, London, 2009 362
G7 Summit, 1991 288
Gabon 224, 228
Gaia Amazonas 200

GDP (gross domestic product) 136, 139, 175
genocide 118, 158, 223–4, 228
geographical isolation 80–1
Germany 339, 346, 370, 382, 396
germination, role for animals 69
Ghana 236–7, 244, 247–52, 260
Giam Siak Kecil Biosphere Reserve 335, 338
glaciations / Ice Ages 38, 56, 78
glaciers, melting 302
glasswing butterfly 95, 96
Global Canopy Programme 355
Global Forest Watch 383–6
global warming
carbon dioxide and 39
mosquitoes and 326
record temperatures, 2014–2016 365
sea surface temperatures 45, 51–2
2° warming target 42–3, 358
see also climate change
Global Witness 292
Gold Coast see Ghana
Gondwanaland 87, 208
Google LLC 384
Gore, Al 368
gorillas
lowland 213, 230, 238
mountain 228, *229*
threats to 226, 228, 240–1
Gorshkov, Viktor 25–7, 48–9
governance, questions of 250, 293, 388, 394
GPS (global positioning systems) 384, 385
Great American Biotic Interchange 77
Great Barrier Reef 49, 52, 293
Great Plains 11, 30, 141
Greenbury, Aida 332–5
greenhouse gases 41, 286
see also carbon dioxide
Greenpeace
campaigning on beef 169
campaigning on cocoa 385
campaigning on mahogany 163
campaigning on palm oil 314–15
campaigning on paper 312–14, 331–2
campaigning on soya 165–8
campaigning on temperate forests 208
strategy toward multinationals 345
technological resources 386
Guatemala 108, 127–8, 130
Guinea, New *see* New Guinea; Papua New Guinea

Guinea forests, West Africa 101, 118
  Lower Guinea rainforest 224, 234, 238–9
  Upper Guinea rainforest 213, 234, 246
Gulf Stream 214
Guzman, Abimael 185

H

Hadley cells / Hadley circulation 26, 28–9
Haiti 106, 144–7
'hamburger connection' 130
Hardenberg, Walter 120
Hareide, Dag 201
Harrods protests 154, 155, 156
Hawaii
  colonizing species 323, 325–6
  honeycreepers 67, 323–4
  human settlement 325–6
  remotest rainforests 322
health effects of droughts 140
The Heart of Darkness, by Joseph Conrad 224
heat stress and coral bleaching 53
heatwaves 39, 45
Heliconia 67
helicopters, reaction to 121, 123
Hemming, John 60–2, 72, 116, 122, 203
herbicide-resistance 141
herbivores, predation of 73
'herring-bone' deforestation 133, 376, 377
highland rainforests see upland
Hispaniola 106, 131, 144–7, 156, 282
Hoh Rainforest 204–8, 211, 214
Hollande, François 367, 369
Honduran tent-making bats 71, 72
Honduras 107, 128, 181
honeycreepers 67, 323–7
hornbills 74, 256, 318–19, 321, 328–30, 388
  seed dispersal by 68
hotspots
  biodiversity 99–102, 149, 152, 327
  deforestation 227
howler monkeys 75
Humboldt, Alexander von 16
Humboldt Current 210
Humboldt's Law 88
hummingbirds
  in Asháninka territory 184

inspiring pigment innovation 96
migrations 211–12
Piedras Blancas National Park 86
as pollinators 67
possible convergent evolution 75
hunter-gatherer lifestyles 111, 230–1, 234, 294
hunting
  ban, Costa Rica 130
  'bushmeat' 196, 226, 241
  Central Africa 231–2
  compounding fragmentation effects 152
  by indigenous people 114–15
  and poverty 269
  using snares 318–19
  West Africa 240
Hurricane Matthew, 2016 144, 145
hurricanes and tornadoes 49, 144, 151
hydroelectricity
  and conservation economics 174, 176
  and drought risk 139, 172, 258–62
  and soil erosion 260
hydrological cycle see water

I

ibis 84, 147
Ice Ages 38, 56, 78
ice cores 37–8, 53, 222
ICOM Cooperative 264–5
i'iwi (bird) 324
illegal burning 300
illegal hunting see poaching
illegal logging
  Amazon basin 133, 202
  attitude of rainforest countries 208, 346, 363, 370–1
  Congo Basin 226
  Global Forest Watch and 385
  illegally harvested mahogany 154–6, 158, 163–4
  Indonesia 290, 302–6, 309, 312, 333–6
  potential for fines 394
  threat to indigenous peoples 190
  timber smuggling and 248
  West Africa 238, 247, 266
IMF (International Monetary Fund) 251–2, 306–10, 345
Incas 110, 117, 184
India
  Periyar tiger reserve 328, *329,* 388, *389*
  Western Ghats 35, *44,* 327–30, 388
indigenous people, generally
  agricultural practices 108
  Amazon basin 10, 118

brutality toward 115–16, 121, 202, 293
Central Africa 230–3
conflict with 61
conflict with logging 231–2
conflict with mining companies 202
crops domesticated by 108
deforestation threat 124
disease susceptibility 116–17
education 195
empowerment 189, 199–200, 394
enslavement 116–21, 123, 184, 200, 203, 221–3, 234–7, 293
improving nutrition 194–5
involvement in conservation 195–7, 388
land rights 185, 201–3, 232
of New Guinea 294–6
plantations by 113
as resettlement victims 289–91
social organization 202
surviving groups 122–4
voluntary isolation of 121
indigenous peoples, named groups
Asháninka 122, 182–99, 340, *384*, 395
Awajún 199
Ayoreo 122
Ba'Aka 230, *231*
Cinta Larga Indians 121
Jirrbal tribes 293
Olmec Indians 108
Paraná people 61
Penan people 287–9
the Tupi 118, 122
indigenous populations
African and Latin American contrasted 232
Asian and Latin American contrasted 279
collapse, Amazon basin 118
Indochina, forests 327
Indonesia
carbon storage and carbon emissions 41
coal and oil extraction 303
colonizing species 78–9
drought of 2015 300–3, *309*
forest and peatland fires 298, 300–3, *308*, 377
formation of peat swamps 41
map 276–7
medieval importance 279
recent rainforest losses 297
and 'The Coral Triangle' 51
transmigration policy 289–91, 306–7
see also Borneo; Java; Papua; Sumatra
Industrial Revolution, fourth 374
infant mortality 192–3
influenza 116, 157, 193
insects as pollinators 65

Institute for Sustainability Leadership 38
inter-tropical convergence zone 45
interglacial periods 38, 78
International Conservation Union 386
iridescent colors 94, 96, 281
island biogeography theory 150, 152
'island effect' 150
islands, rapid evolution on 323
isolated animal populations 151
Ivory Coast / Côte d'Ivoire 236–9, 242–5, 254–9, 263–4, 271–4, 377
see also Tai Forest National Park

J

Jakarta 297–8, 307
Jamaica 131, 147, 148, 156
Jambi province, Sumatra 290, 302, 304, 306, 340
Jans, Beat 353
Japan 41, 97, 204, 283, 382
Java sparrows 325
Javan rhinoceros 319
Javari Valley 202–3
Jenkins, Dilwyn 184
Jurassic forests 76

K

Kalimantan, East and West 299, 322, 339
Kariba Dam 261
Kenya 61, 370
Kerry, John 368
Kew Gardens 97–8, 119
keystone species 75, 336
Khmer Empire 278–9
kingfishers 97
Kivu, Lake 226
Kohler, Marcus 310
Konon, Adrien 268
Korup National Park 238
Kuahara, Chris 199

L

La Selva Biological Station 69, 178
Lake Chad 56
Lake Kivu 226
land bridge between the Americas 77, 81, 215
land ownership disparities 128, 130

land rights
  Asháninka people 185
  Central Africa 232
  conflict with development 233
  indigenous people 185, 201–3, 232–3
  Papua New Guinea 294
  resettlement and 290
landlessness driving forest incursions 128, 338
landslides and mudslides 147, 187
languages, local 188, 198, 201, 230, 294
Laos 283, 292
Latin America, hotspots 102
Lavazza 340
Lawrence, Deborah 30–2, 36, 142
leaf eating by animals 65
leaf litter decomposition 58, 59
Lees, Andrew 9–10, 158, 230, 247–8
Leopold II 223
Leuser landscape, Sumatra 321
Liberia 237–8, 370, 391–2
lichens 33, 88, 205–6, 214
life expectancy 180–1, 192
Lifescaped company 96
logging
  Caribbean vulnerability 131
  concessions 305–6
  and conflict 237–8
  Congo Basin vulnerability 226, 232
  as first stage of forest degradation 47–8, 131, 156, 188, 226, 232, 284–5, 305
  incursions by loggers 186, 198–9
  Indonesian deforestation 283–4, 288
  in Sumatra 303
  see also illegal logging; timber companies
Los Angeles Cloud Forest 33, 93
'lost cities' 106, 110–15, 278
Lower Guinea rainforest 224, 234, 238–9
Lula, President (Luiz Inácio Lula da Silva) 362
Lutzenberger, Jose 157

M

MacArthur, Robert 150, 152
macaws 70, 147, 148, 190
Madagascar 230, 370
mahogany
  CITES listing 160–4
  illegal harvesting 154, 156–8, 190
  protection 170, 202
  qualities and desirability 154–6, *162*

Mahogany is Murder campaign 154, 155, 157, 161
maize (corn) 98, 102, 109, 112, 139, 242
  corn ethanol and corn syrup 141–4
Major, John 288
Makarieva, Anastassia 25–7, 48–9
mammals, as seed dispersers 68
mangrove forests
  Piedras Blancas National Park *82, 83*
  protective effects 51, 146
manioc 108, 188, 191, 193
Manser, Bruno 288
Manu National Park, Peru 395
maps
  rainforest areas 104–5, 218–19, 276–7
  showing concessions 337–8, 340
Maraca Project 60, 62
Marco Polo 279
marine forests 83–4
Mars, Inc. 267, 375, 381
marsupials 87
Mason, Richard 60–1
mass extinctions 52, 149
massacres 121, 202
Mattel Inc. 312–14, 331
Matthew, Ed 311
May, Brendan 332
McCann Erikson 158
McDonald's 167, 168, 381
McLaren, Duncan 352, 353
medicines
  animal parts 241
  Chinese medicine 319–20
  rainforests as sources 97–9, 115
Mekong river 292
Melanophila beetles 96
Merkel, Angela 362
Michelsen, Alfonso López 200
Mighty Earth group 236
migration
  of birds 84, 211–14
  of people to Europe 242, 257, 265
  reducing by poverty reduction 398
mining activities
  conflict with indigenous people 202
  Congo Basin 228
  Madagascar 230
missionaries 122, 184, 200, 223, 293
Mitchell, Andrew 355, 366
Mohamad, Mahathir bin 288
Monbiot, George 157
Mondelēz International Inc. 263–4,

266–7, 270, 381
see also Cocoa Life initiative
money laundering 127
mosquitoes 140, 326
moths 65, 79, 194
muchachos 118
mudslides and landslides 147, 187
multinational companies
enlightened policies 263–6, 354, 381,
394
implicated in deforestation 10, 228,
310–11, 315, 345
power shift toward 345
zero deforestation commitment 381–
282
murder of environmentalists 128, 288
murrulets, marbled 206
Musi, river 298–9
Myanmar 292, 327
mycorrhizae 62–3, 73

N

narco-cultivation 127, 133, 185–6,
189–90, 195
National Investigations Committee
(Ghana) 247–50
National Parks and reserves
Amani Forest nature reserve 380–1
Berbak and Sembilang National Parks
340
in the Caribbean 131
Chiribiquete National Park 396
contribution to forest preservation 203,
378, 387–8
in Costa Rica 172, 174, 180
Giam Siak Kecil Biosphere Reserve 335,
338
illegal clearance in 333, *334, 336*
Korup National Park and Cross River
Park 238
Olympic National Park 205–7, 211, 215
Otishi and Manu National Parks 395
Panama–Costa Rica cross-border 131,
178
Piedras Blancas National Park *82,* 83, 85
resourcing problems 389
Tambopata-Candamo reserve *17*
Tanjung Puting, Borneo *322*
Tawau Hills National Park 82
Virunga National Park 228
Volcanoes National Park, Rwanda *229*
see also Tai Forest
Ndiaye, Mbalo 270–1

Nestlé 267, 331, 366, 381
net positive goals 387
Netherlands 163, 280
New Guinea
Araucaria trees 76
hunter-gatherer survivals 294–6
origins of bananas 283
origins of sugar cane 282
political division 294
West Papua 294, 295
see also Papua New Guinea
New Tribes Mission 122
New York Declaration on Forests 386
Nicaragua 77, 122, 128–30, 181
Nigeria 224, 236, 238–9, 258
nitrous oxide 286
Nobre, Antonio Donanto 140
nocturnal animals 87
Norway 204, 339, 362, 370, 378, 382,
391–2, 396
nutrients
held primarily in plants 58–9
from Saharan dust 56
soil enrichment 111, 194
nutrient cycle / recycling 59, 61–3,
266–7
'nutrient knife edge' 56–62

O

Oahu amakihi (bird) 323, 327
oaks, evergreen 88
Obama, Barack 364–5
ocean acidification 51–2
ocean circulations 78
ocean temperatures see sea surface
ocelots 88, 92, 153, 196
OECD (Organization for Economic
Co-operation and Development) 307
ohi'a trees 324, 326–7
oil palms see palm oil
Olympic National Park 205–7, 211,
215
Hoh Rainforest 204–8, 211, 214
opossums 77, 87
orangutan 314, 321, 322
Orellana, Francisco de 110
organic material, recycling 58
Orinoco River 16, 132–4
Oroxylum trees 67
Otishi National Park 395
Overseas Development Administration,
UK 252

Owen, Matthew 182, 186, 188–9, 192, 195, 197–8
oxygen, from photosynthesis 18–19

P

Pacific Northwest 204–8, 211
packaging, sources 313, 381
paint research 94, 96
palm oil / oil palms
  concessions 305
  deforestation in Guatemala 127
  deforestation in the Amazon basin 133
  fruit *284*
  illegal plantations 333, *334*
  Malaysia and Indonesia 283–91, 298–303, 305–6
  overproduction and price collapse 308–10
  pressure on customers 314–15, 381
  West African rainforest 236–7, 243, 274
  yield and applications 285–6
palms, rainforest 113–15
Pan-American highway 77
Panama 130, 178
Panama Canal 150, 152
Pangaea 76
pangolins 73, 319–20, *320*
paper companies 305, 310–14
Papua, West 294, 295
Papua New Guinea 295
  Cool Earth in 189
  land rights 294
  leadership over conservation 362
Páramo 88
Paris, Climate Change Conference, 2015 337, 365, 366–71, 386, 394
Parker, Andrew 95–6
parrots 9, 60, 69–70, 84, 134, 184, 187
Patagonia 77, 208–9
'Path of the Panther' project 178
pathogens 98, 116, 151, 325–6
peat swamps
  burning 45, 298–301, 377
  formation beneath rainforest 40–1, 83, 335, *338*
  re-wetting 336–7
Peña, Jaime 195–7, 384
Penan people of Borneo 287–9
Periyar Tiger Reserve 328, 329, 388, 389
Peru

Asháninka people 122, 182–99, 340, *384,* 395
  deforestation commitment 370
  Tambopata-Candamo reserve *17*
Peruvian Corporation 184
pesticides 181, 240, 286
pet trade 241, 325
Philippines 48, 51, 102, 278, 280, 283–4, 291, 308, 327
phosphorus uptake 63
photosynthesis
  driving the rainforest water cycle 15–16
  releasing oxygen 18–19
Piedras Blancas National Park 82, 83, 85
pigs
  eaten by forest peoples 113, 191, 240
  predators of 73, 318
  seed dispersion by 71, 74
pineapples 109, 171, 176, 188
pit vipers 92
plants
  communication via roots 63
  decomposition and peat swamps 40, 83
  growth rate and CO_2 levels 43
  medicinal 97
  sexual reproduction 64, 72
  variety in Costa Rica 81
  water pressure inside 18
plant species
  biodiversity hotspots 99
  Heliconia 67
plant toxins
  accumulation by caterpillars 65, 67
plantations
  adverse effect on soil moisture 27
  by pre-Columbian farmers 113
  see also palm; pulpwood; soya
poaching
  fragmentation effects 152
  trade in wildlife 319–20
  West Africa 238, 239–42
politics
  Brazilian inconsistency 378
  emergence of climate concerns 353–4
  Indonesian transmigration policy 290
  power shift toward multinationals 345
  REDD unintended consequences 358
pollen, cloud seeding by 22
pollination
  bats as pollinators 67
  birds as pollinators 67, 326
  crops by rainforest species 99
  and ecological interconnections 74

insects as pollinators 65
wind pollination of grasses 64
populations
  Africa and Latin America contrasted 232
  Congo Basin 224, 230–1
  of indigenous groups today 118
  supported in pre-Columbian America
    106, 115
population crashes in rainforest
  countries 10, 118, 223
population growth
  India 329
  and overall environmental change 262
  rainforest countries 12, 242–3, 254–6
port facilities 135–7, 164–6, 237, 265
Portugal
  African slave trade 221
  colonization of Brazil 117–18
positive feedback 45–6
potatoes 102, 109
poverty 74, 242, 245, 265
  and corruption 242, 247, 306
predators
  of cloud forest frogs 92
  controlling herbivores 73
  convergent evolution 75
  of marine forests 84
  minimum viable populations 151
Prescott, Tom 97–8
primary forest
  defined 85
  loss of, in Sumatra 291
The Prince of Wales, HRH 267, 359–
  62, 367, 396
The Prince's Rainforest Project 359–63
privatization 309, 346
Public Eye on Davos campaign 351–2
public spending cuts 252, 309, 345,
  378
Puerto Rico 131
Pulgar-Vidal, Manuel 370
pulp mill, Parawang, Sumatra 311
pulpwood plantations 164, 291, 298–
  300, 306, 311–12, 316, 333, 337
  see also paper
The Putumayo, the devil's paradise . . .
  by Walter Hardenberg 120

Q

quetzals 89, 90
Quinchori, Saul 194

R

Raiders of the Rainforest (television
  documentary) 251
rain
  effect of deforestation 18, 28, 30
  effect of forest fires 302
  see also droughts
rainforest, temperate 59, 81, 204
rainforest, tropical, use of the term
  16–17
Rainforest Action Network 163, 347,
  386
rainforest countries
  economic collapse 308
  London meeting, 2009 362
  Paris summit, 2015 367–8, 370–1
  population crashes 10, 118, 223
  population growth 12, 242–3, 254–6
  resistance to forest protection measures
    208, 346, 363
  resistance to indigenous land rights
    232–3, 287, 289
Rainforest Foundation 201, 247
rainforest loss see deforestation
rainforests
  as carbon stores 37, 39–42
  connections between temperate and
    tropical 211–12
  definition and use of the term 16–17
  as a 'green ocean' 21–3
  neighboring deserts 27
  non-indigenous inhabitants 123
  nutrient and hydrological cycles 61–2
  peat underlying 40–1, 83, 335, 338
  resilience in drought conditions 46
  restoration targets 370
  role in global climate 30, 43–4
  role in weather systems 23–8
  as sources of medicines 97–9
  types of 81, 92–3, 211–12
  unsuitability for farming 58
  upland rainforests 82–3, 87–91, 100,
    101
  see also deforestation; forests
ramins 284, 312
Rawlings, Jerry 247–8
red-eyed tree frog 91–2, 93
REDD (reducing emissions from
  deforestation and degradation) 271,
  356–8, 369–71, 382, 391–3
redwoods, California (Sequoia) 204–5,
  209
renewable energy 175–6, 282, 390,
  394

reserves see National Parks
resettlement programs 124, 132–3, 289–91
Riau province, Sumatra 40, 291, 306, 308, 311, 316, 332, 334, 335
rice, rainforest origins 283
Rice, Tim 250
Richards, Paul 17
Richards, Steve 295
Rio Aripuanã 121
Rio de la Plata basin 29, 139
Rio Ené 182, 185–6, 187, 188, 190
Rio Esquinas 83
Rio Negro 16, 118
Rio Perené 186
Rio summit see Earth Summit
Rio Tapajos 20, 110, 132, 164, 165
Rio Tinto 230
rivers, named
  Mekong river 292
  Orinoco 16, 132–4
  river Kaveri 330
  river Musi 298–9
  Sassandra river 258, *259*
  Ubangi river 225
  Volta river 248, 260
  Zambezi river 261
  see also Amazon; Rio
roads
  forest fragmentation 47, 132, 136, 227–8, *376, 377*
  highways, BR163 and BR364 133, 164
  Pan-American highway 77
  threat to indigenous communities 198–9
Robertsbridge advisory group 331–2
rodents 18, 72–3, 100, 152–3
Rodriguez, Carlos Manuel 172–6
Roosevelt, Theodore 125, 127
root systems 62–3
Roy, Assie 246
Royale, Segolene 368
rubber
  Amazon deforestation and 132
  Congo exploitation for 223–4
  history of exploitation 118–22, 184, 200
  Ivory Coast 243
Rwanda, Rift Valley 102

S

Saharan dust 55–6
saltwater forests 83–4

San Gerardo Valley 88
sandpipers 84, 189, 213–14
Sankuru Nature Reserve 227
Santarém 136, 164, 166–7
Santos, Juan Manuel 396
São Paulo 29, 126–7, 136, 138–41
Sargent, Randy 375
Sarkozy, Nicolas 362
Sassandra river 258, 259
satellite data
  Earth Time Lapse sequence 372, 375–7
  Landsat and Sentinel data 383
  monitoring forest fires 302
  monitoring forest fragmentation 47, 133
  monitoring forest losses 11, 42, 227, 286, 383
  monitoring Saharan dust 56
  monitoring soya production 169
  revealing depopulated regions 222
  revealing 'lost cities' 110
Sauven, John 166–7, 169, 314–15
Schimper, Andreas 17
sea forests / marine forests 83–4
sea level changes 39, 78–9
sea surface temperatures 45, 51–2
seasonal change and its absence 19, 80
Seattle, WTO Ministerial Conference, 1999 346–9
Secrett, Charles 12, 332
sediments, water-borne 50–1
seed dispersion
  by animals 66, 68–74, 152–3, 177
  by birds 66, 68, 177, 318
self-cleaning surfaces 96–7
Selva Bananito project 176–8
La Selva Biological Station 69, 178
Sembilang National Park 340
Sendero Luminoso ('Shining Path') 185–6
sexual reproduction in plants 64, 72
Sheppard, William Henry 223–4
Sierra Leone 48, 237–8
Sinchicama, Marcial 183
'sky rivers' 28–32, 139
the slave triangle 222
slavery
  abolition 236
  descendants of slaves 123, 203
  and forest extent 221–3
  indigenous people in Africa 221–3, 234–7
  indigenous people in Australia 293
  indigenous people in South America 116–21, 184, 200

Sierra Leone conflict 237
sloths 71, 210
smallholders
  contribution to deforestation 137
  global importance 254
  government support 272
  improving productivity 267, *269*, 340,
  391, 394
  indigenous people as 193
  insecurity 256, 265
snake skins as inspiration 96–7
snakes, in Costa Rica 81, 92
Soco International plc 228
soil degradation and carbon emissions
39
soil erosion 50–1, 113, 260
soil improvement
  indigenous peoples 194
  pre-Columbian civilizations 110–15
  under tall trees 266, 341
soils
  as ecosystems 62
  nutrient depletion 57–8
solar panels 95, 96
solar radiation, reflection 35–6
Soubré dam 258–9
South Dakota 141–3
South Sumatra 298, 306, 339–40
soya 126, 133–6, 139, 141, 360, 366,
373, 381
The Soya Rush 164–70
species
  discovery of novel species 100, 102,
  295–6, 321
  diversity in rainforests 81, 93, 100, 149
  number identified 79–80
spores, fungal 22, 73
Stern, Nick 368
storm surges 51, 146–7
strangler figs 33
structural adjustment programs 251–3,
381
sugar
  domestication 282
  from photosynthesis 18, 62
  slavery and 221
  sugar cane 126, 130–1, 282, 293
Sulawesi 68, 289, 327
Sumatra
  APP target landscapes 331–5, 339
  as a biodiversity hotspot 327
  deforestation and forest fragmentation
  291, 316, *317*, 318–19, 337–41
  emissions from burning peat 45, 302,
  377

illegal logging *303*, 304–6
indigenous people 289–91
Leuser landscape 321
logging camp *303*
Marco Polo on 279
oil palm plantations 286
peat swamps *40*, 41
pulpwood concessions *308*, 310, *311*
sandpiper migration 213
transmigration to 290
Sumatran elephants 318–20, 321, 336,
339–40
Sumatran rhinoceros 319, 321
Sumatran tigers 303–5, 317–20, 334,
336, 339–40
sunbirds 75, 316
Sundaland 102, 327
sunlight, competition for 60, 61
Survival International 121, 202
Sushisamba restaurant 197
swifts 193, 212–13, 373
Syria, drought and conflict 375

T

Tai Forest National Park 241, 244,
249
  and cocoa growing 272
  pressures on 245–7, 255, 263, 270,
  375–7, 389
  role of elephants 68
Talamanca Mountains, Costa Rica 88,
100
tanagers 71, 212
Tanzania
  Amani Forest 380–1
  Usambara Mountains 78, 380
  WCMC conservation work 100
tapirs 68, 73–4, 88, 150, 184, 196,
304, 340
Taroveni, Peru 190–5
Tawau Hills National Park 82
tax evasion 250
Taylor, Charles 237–8
tea, rainforest origins 281–2
Temer, President Michel 137, 202
temperate forests 59, 81, 204
temperature change with altitude 88
teosinte 109
termite mounds 97
terra preta soils 111–12
Tesco 313–14
Tesso Nilo 332, 334, 335

Thailand 32, 278, 282–3, 291–2, 308, 310
Third World debt / debt crisis 10, 251–2, 308–9
'three sisters' crops 112
Tianjin, World Economic Forum, 2016 372–7, 379
tigers
  Periyar Tiger Reserve 328, 388, *389*
  population recovery 389
  Sumatran tigers 303–5, 317–20, 336
  Western Ghats 328
timber certification schemes 160
timber companies
  conflict with indigenous people 157
  EU timber imports 10, 292, 385
  export of unprocessed logs 309
  reaction to Mahogany is Murder 158–60
  see also logging
timber smuggling 248
Togo 234, 260, 370
tomatoes 109
toucans 34, 60, 68, 70, 74
tourism
  Costa Rica 171–2, 174–9, 181, 393
  ecotourism 174, 178, 181, 393–4
  Western Ghats 388
toxin accumulation by caterpillars 65, 67
trade winds 27
transmigration policy, Indonesia 289–91, 306
transparency and accountability 383–6
transpiration 18, 22, 27
trees
  emergent trees 68, 82, 246, *249*
  exceptional size and age 205, 209
  fungus disease, Hawaii 326
  relationships with animals 72
  small, on branches of large 86
  small, on fallen logs 205
  stream formation 35
  water transpired by 21
tree ferns 35, 76
tree species
  alerce 209
  Araucaria trees 76
  Beilschmiedia miersii 210
  Brazil nut trees 70
  Cecropia 71
  Daniella 234
  diversity in rainforests 81, 208
  evergreen oaks 88
  lobelias 324–5
  ohi'a 324, 326–7

Oroxylum 67
ramins 284, 312
sequoia 204–5, 209
Sitka spruce *207*
yellow meranti 82
see also mahogany
trogons 89, 177, 179
Tropical Forest Resources Assessment Project (UN) 11
Tropical Forestry Action Plan 358
Tropical Forestry Alliance 379
The Tropical Rain Forest, by Paul Richards 17
the Tupi 118, 122
2° warming target 42–3
Ty Canol woodland 214–16

U

Uganda, Rift Valley 102
UK see Britain
'uncontacted tribes' 121–2, 202–3
Unilever 314–15, 366, 381
United Cacao 385
United Nations
  aircraft emissions 'mechanism' 391, 393
  Climate Change Conference, Bali, 2007 356–9, 362
  Climate Change Conference, Copenhagen, 2009 362, 363–6, 394
  Climate Change Conference, Paris, 2015 337, 365, 366–71, 386, 394
  declaration of biosphere reserves 335
  Food and Agriculture Organization 137, 258
  New York Declaration on Forests 386
  Tropical Forest Resources Assessment Project 11
  World Conservation Monitoring Centre 100, 252
  see also REDD
United States
  Great Plains 11, 30, 141–4
  Olympic National Park 205–7, 211, 215
Unzui, Amani Kwasi 257
upland rainforests 82–3
  Costa Rica 87–91, *101*
  Tanzania 100
Upper Guinea rainforest 213, 234, 246
urbanization see cities
Usambara Mountains, Tanzania 78, 380

V

Valdivian rainforests 208
Value of Nature presentation 375
Vandecar, Karen 30–2, 142
varzea flooding 20, 113
Venezuela 48, 132–3, 395
Vietnam 281, 283, 292
violence
 against environmentalists 128, 288, 347
 against indigenous people 157, 202, 290
Virunga National Park 228
Volta river 248, 260
von Hildebrand, Martin 200–3, 395–6
von Humboldt, Alexander see
Humboldt

W

Wales, Ty Canol woodland 214–16
Wallace, Alfred Russel 283
Wallace's Line 327
warfare see conflicts
Washington Consensus 345–6
water
 extraction in cloud forests 33
 long-distance movement 28–32
water cycle 15, 22, 61–2
water security 12, 32, 262, 329, 361,
 387, 395, 397
 see also droughts
water transfer, tipping points 32
wealth inequality 136
weaverbirds 239–40, 256
wellbeing measures 180–1
Wen Jiabao 365
Western Ghats, India 35, 44, 327–30,
 388
 Periyar Tiger Reserve 328, 329, 388,
 389
Wickham, Sir Henry 119
wild fires see forest fires
wildlife
 assessment within concessions 336
 defaunation 73–4
 ecotourism 174, 178, 181, 393–4
 endangered, South America 209
 extinctions, Madagascar 230
 extinctions, New World 147–53
 illegal trade in 319–20
 as proportion of vertebrate biomass 373
 REDD concerns 358

surveillance tools 285
 threat from alien species 322–7
 threats to, in 1990 10
 Western Ghats 328
 see also biodiversity; hunting
Wilson, Edward 150, 152
wind pollination, grasses 64
wood warblers 211–13, 215
'wood-wide web' 62–3
woodland see temperate forests
woodpecker, black-cheeked 66
World Agroforestry Centre 340
World Bank
 actively combating deforestation 382
 Amazon deforestation and 132–3
 Friends of the Earth objectives 10
 funding, Ivory Coast 272
 Ghanaian economic reforms 252
 indirectly promoting deforestation
 306–10
 in Indonesia 339
 and the Washington Consensus 345
World Conservation Monitoring
Centre, UN 100, 252
World Economic Forum
 Davos, 2001 349–53
 Tianjin, 2016 372–7, 379
World Heritage Sites, Unesco 228
World in Action (television series) 251
World Resources Institute (WRI) 51,
 227, 383
World Trade Organization (WTO)
 307, 346–7, 349
 Ministerial Conference, Seattle, 1999
 346–9
WWF (World Wildlife Fund) 160,
 228, 386, 389, 398

Y

yellow meranti 82
Yilditz, Kerim 353
Yorke, Thom 363, 364
Yudhoyono, Susilo Bambang 357, 362

Z

Zambezi river 261
Zambia 261–2
zero deforestation companies 381–2,
 397
Zimbabwe 261

COVER PHOTOS

FRONT COVER: Danum Valley, Sabah, Borneo. BACK COVER: Silverback gorilla, Volcanoes National Park, Rwanda. FRONT ENDPAPER: Macaws, Tambopata Candamo Reserve, Peru. All images © Thomas Marent

About Island Press

Since 1984, the nonprofit organization Island Press has been stimulating, shaping, and communicating ideas that are essential for solving environmental problems worldwide. With more than 1,000 titles in print and some 30 new releases each year, we are the nation's leading publisher on environmental issues. We identify innovative thinkers and emerging trends in the environmental field. We work with world-renowned experts and authors to develop cross-disciplinary solutions to environmental challenges.

Island Press designs and executes educational campaigns, in conjunction with our authors, to communicate their critical messages in print, in person, and online using the latest technologies, innovative programs, and the media. Our goal is to reach targeted audiences—scientists, policy makers, environmental advocates, urban planners, the media, and concerned citizens—with information that can be used to create the framework for long-term ecological health and human well-being.

Island Press gratefully acknowledges major support from The Bobolink Foundation, Caldera Foundation, The Curtis and Edith Munson Foundation, The Forrest C. and Frances H. Lattner Foundation, The JPB Foundation, The Kresge Foundation, The Summit Charitable Foundation, Inc., and many other generous organizations and individuals.

The opinions expressed in this book are those of the author(s) and do not necessarily reflect the views of our supporters.

Island Press | Board of Directors

Pamela Murphy
(Chair)

Terry Gamble Boyer
(Vice Chair)
Author

Tony Everett
(Treasurer)
Founder, Hamill, Thursam
   & Everett

Deborah Wiley
(Secretary)
Chair, Wiley Foundation, Inc.

Decker Anstrom
Board of Directors,
Discovery Communications

Melissa Shackleton Dann
Managing Director,
Endurance Consulting

Margot Ernst

Alison Greenberg
Executive Director,
Georgetown Heritage

Marsha Maytum
Principal,
Leddy Maytum Stacy Architects

David Miller
President, Island Press

Georgia Nassikas
Artist

Alison Sant
Co-Founder and Partner,
Studio for Urban Projects

Ron Sims
Former Deputy Secretary,
U.S. Department of Housing
   and Urban Development

Sandra E. Taylor
CEO, Sustainable Business
   International LLC

Anthony A. Williams
CEO & Executive Director,
Federal City Council